Advances in Anatomy, Embryology and Cell Biology
Ergebnisse der Anatomie und Entwicklungsgeschichte
Revues d'anatomie et de morphologie expérimentale

48 · 3

W0106286

Ferdinand Vanpeperstraete

The Cartilaginous Skeleton of the Bronchial Tree

With 42 Figures

Springer-Verlag Berlin Heidelberg New York 1973

Dr. Ferdinand Vanpeperstraete, Agrégé
Department of Anatomy and Comparative Anatomy
State University, Gent Belgium
Ledeganckstraat 35

ISBN 978-3-540-06536-4   ISBN 978-3-642-95252-4 (eBook)
DOI 10.1007/978-3-642-95252-4

Contents

I. Introduction

A review of the publications, dealing with the morphology of the cartilages of the tracheo-bronchial tree, shows how scarce they are and how fragmentary the contributions based on research. Isolated parts only of the bronchial tree have been investigated, mostly in single specimens or small series.

Anatomical textbooks merely state that the trachea and main bronchi are supported by rings and the more distal branches by irregular, circumferentially placed plates which become smaller towards the periphery, until they disappear in the bronchioli.

It is at once obvious that this old-time view is not only superficial, but it leaves one completely ignorant about the site where rings cease and plates begin.

No information is available about the arrangement of cartilages around the bifurcations of the main bronchi and contradictory descriptions are given about the cartilaginous architecture of the lobar bronchi.

A more searching study reveals that cartilages are hardly mentioned in the huge amount of literature on bronchial anatomy which has accumulated since 1880. It is therefore not exagerated to say that the study of this subject has been grossly neglected.

A more accurate knowledge of the cartilaginous anatomy of the bronchial tree would nevertheless be useful practically as well as theoretically: the presence of cartilage or its absence is taken into consideration in bronchial and pulmonary anomalies; obstruction of the bronchi is conditioned by their rigidity, in other words by the shape and situation of cartilages; alterations of the shape of the airways by tuberculous and inflammatory processes are influenced by them; the many difficulties encountered by surgeons in closing the bronchi after pulmonary resections are largely due to the spring-like action of the rings and in the operation of resection-anastomosis of the main bronchi, the surgeon fails to realize how he alters the bronchial skeleton; bronchi surrounded by rings will respond physiologically in a different way than bronchi surrounded by plates.

Our first purpose in undertaking this study was to establish exactly in which bronchi rings were found and in which plates and determine where the former ended and the latter began.

Consequently the unexplored cartilages of the secondary bifurcations and the lobar bronchi were studied first. But gradually and inevitably the main bronchi, the distal trachea and its bifurcation and the proximal part of the segmental bronchi became involved and our investigation expanded to become a comprehensive study of the whole cartilaginous skeleton of the bronchial tree. A precise answer could then be given to a number of questions which had remained vague or controversial and our second purpose met.

Our observations, descriptions and conclusions were based on a large series of preparations. The ambition of meeting a flagrant need for a systematic study could thus be fulfilled.

By extending our investigations to the organogenetic appearance of cartilage in human foetusses and their morphology in some animal species, we were able to venture an explanation of the phylogeny of the bronchial tree and the morphogenetic mechanism which determines the shape of the cartilages.

To stimulate new interest in the subject was our final aim.

II. Historical Review

A. From Cl. Galenus to Chr. Aeby (1880)

In the seventh book on the use of the parts of the body, Περὶ χρείας μορίων, Cl. Galenus (2nd Century) remarked that what was called τραχεῖα ἀρτηρία by some, βρόγχος by others, contained a great deal of χόνδρος. This cartilage is divided into rings which lie one above the other, are joined by strong bands and connect the larynx to the lungs. As no cartilage is present where the trachea lies on the oesophagus rings are not complete but resemble the old Greek sigma or "C" and are therefore sometimes called "sigmoid". Cartilages are also sigmoid shaped in all the branches of the airways (Ch. Daremberg, 1854).

A. Vesalius (1543) agreed with this description and illustrated it as far as the trachea was concerned. But C. Bauhinus (1621) observed that as soon as the main bronchi enter the lungs and split into lobar ones, the "cartilagines, nunc triangulares, nunc quadratae, nunc alio modo formatae, per pulmonum ad extremum disseminantur" and he corrected Vesalius' sketch accordingly.

The erroneous concept of complete rings was introduced when F. Ruysch (1665) illustrated the bronchial arteries which he claimed he was the first to describe, on the posterior aspect of a bronchial tree, the tracheal rings of which were shown to almost touch each-other.

To the charge of heresy of which G. Bidloo (1697) accused him, he replied in his "Responsio" (1697) that the preparation did not belong to man, but to a newborn calf, the tissues of which were soft and had possibly shrunk in the process of fixation that the artist had not been very accurate and that the illustration was meant after all to show the bronchial arteries and not the cartilages.

Nevertheless did Th. Willis (1674) also describe the rings as "circulares" in the main bronchi, followed by Ph. Verheyen (1711), J. Palfyn in the Dutch edition of the "Heelkonstige Ontleeding" (1718) and later by J. Lieutaud (1777) and others.

I. de Diemerbroeck (1685) and a few others remained faithful to the views of Cl. Galenus and C. Bauhinus and in 1706 J. B. Morgagni showed convincingly that the tracheal and main bronchial rings were undoubtedly "semi-circulares".

Unanimity was therefore reached no earlier than at the end of the 18th century when time was ripe for new observations and F. R. Buisson remarked in X. Bichat's "Traité d'Anatomie" (1803) that the last tracheal ring was to be distinguished from the preceding ones by its shape which was adapted to the contour of the main bronchi and he added that in many cases the last tracheal ring was fused to one of the main bronchial ones.

The average number of rings in the main bronchi was recorded by J. F. Meckel (1820) as eight in the right and nine to twelve in the left one.

To a prize-essay issued by the German Academy of Sciences in Berlin, F. D. Reisseisen (1822) answered that cartilages are no longer seen in bronchioli half a line (1 mm) in diameter and less and at the same time he described the bifurcation cartilages which surround the branchings of the airways inside the lungs. According to him, they were circular proximally and crescent shaped more distally until they finally disappeared. W. E. Horner (1839) was the first to call attention to their peculiar shape and J. King (1840) showed a number of illustrations, adding that a considerable number of varieties will be met with.

How the morphology of the cartilages is adapted to the shape of the airways in different species, outside as well as inside the lungs, was studied simultaneously in Germany by J. F. Meckel (1833) and in France by G. Cuvier (1840). They accumulated a large amount of new information which now and then differed from one edition to the other. They made the distinction between bony, cartilaginous and fibrous cartilage and when the extra-pulmonary airways did contain cartilage and the intra-pulmonary ones did not, an attempt was made at establishing at which level the transition occucred.

H. von Luschka, whose name reminds the "intercalar" or anomalous cartilages (1861) which are sometimes found in the proximal part of the membranous trachea, is also to be credited with the first satisfactory description of the rings supporting the tracheal bifurcation (1863) and the "carina" a name which he introduced. The last tracheal ring, he believed, can be fused to the preceding one; it has a medial projection which indicates the bifurcation; the first bronchial rings on the other hand are fused medially and lend support to the carina.

Using 125 human and 48 animal preparations, R. Heller and H. von Schroetter (1897) undertook a systematic study of the cartilages in this region. They found that a great number of combinations occurred. They called carinas cartilaginous when they were supported by either tracheal or main bronchial rings, membranous when support was lacking and mixed, when only part of the carina contained cartilage.

A few text-books appearing at that time (J. Henle, 1866; Ch. Morel and M. Duval, 1883; J. Hyrtl, 1885) reproduced a drawing showing the shape of rings in part of the trachea, the bifurcation and the root of the main bronchi. Ph. Sappey's (1874) description which goes with it can be translated as follows: "The rings all resemble each-other. Sometimes however they are unlike the normal and take on varied shapes. They often fork at one end, in which case the following ring usually forks at the other end, restoring the parallelism of the tracheal rings. When the latter does not fork, an extra cartilaginous plate develops, reestablishing parallelism again. At other times they approach each-other, meet at acute angles and are partly fused. Some rings are broader in their middle or at some other site. Others are exceptionally broad, while the ones which precede or follow are more or less reduced in size, although they mostly maintain the same configuration. One finally observes, that, compared to each-other, not two of them look exactly alike".

Concerning the intra-pulmonary cartilages this author still wrote: "The cartilaginous rings are divided in several segments which move apart and encircle the whole circumference of the bronchus... they produce this way a complete ring, broken at intervals".

To this first period belongs a schematic but remarkable reproduction of the cartilaginous skeleton of the human bronchial tree which appeared in K. von

Bardeleben's "Handbuch der Anatomie" in the part on the respiratory apparatus by F. Merkel (1902).

B. From Chr. Aeby (1880)

An entirely new orientation was given to the study of broncho-pulmonary anatomy by the publication of Chr. Aeby's remarkable monograph: "Der Bronchialbaum der Säugethiere und des Menschen" (1880). All links with the past were severed, the descriptive method forsaken for a speculative one and an attempt made at establishing a general axial system of branching of the airways which would apply to all mammals.

New impetus was given to the study of the subject which resulted in a number of remarkable contributions by W. Ewart (1889), A. Narath (1892–1901), G. S. Huntington (1920) and others in which, according to the fashion of the days, a discussion of evolutionary theories was the main concern. At the same time the foundations were laid for a better knowledge of the segmental distribution of the bronchi in the lungs.

Every one of these authors based his conclusions mainly on casts or corrosion preparations which had been made popular again by J. Hyrtl in 1873. They were mistakenly called "naked" bronchi, because the bronchial wall and the cartilages had disappeared in the process of corrosion and what was left was not a bronchial tree but a replica of its inside, its lumen. Consequently the distinction between extra- and intra-pulmonary airways gradually waned and no notice was taken any more of the bronchial wall and the cartilages it contained.

None of the above-mentioned authors took any notice of the morphology of the tracheo-bronchial cartilaginous skeleton, nor mentioned it in his works and it is to be regretted for the sake of our subject that Chr. Aeby (1880) and A. Narath (1901) both illustrated their remarkable monographs with bronchial trees, the cartilages of which were shown to have the same shape in the trachea, the main bronchi and the peripheral airways.

When D. J. Davis (1929) opened the phenomenal series of investigations which would finally culminate in a precise knowledge of the segmental bronchi, their way of branching, their distribution within the lungs and their variations (H. P. Nelson, 1931; R. Kramer and A. Glass, 1932; M. Lucien and P. Weber, 1933; E. Huizinga, 1937; J. H. Neil et al., 1937; E. D. Churchill and R. Belsey, 1939; L. C. Jackson and J. F. Huber, 1943; A. F. Foster Carter, 1944; R. C. Brock, 1947; E. A. Boyden, 1955; to mention only a few of the most important authors and one of their publications) methods of investigation reproducing the inside of the bronchi were almost exclusively used: casts, injection preparations, bronchoscopies (J. Killian, 1902), bronchographies (J. A. Sicard and J. Forrestier, 1928) and the study of the bronchial wall was relegated to the realm of histology.

This is why only a few scattered papers on the morphology of tracheo-bronchial cartilages appeared since Chr. Aeby and why one had to wait until 1937 when W. S. Miller who had studied the work of Th. Willys (1922) discovered that no satisfactory answer had yet been given to the question how bronchial bifurcations are supported by cartilage. He therefore started a series of reconstructions intending to include a series of vertebrates. But he did not get beyond the reconstruction of part of a human bronchial tree and of a guinea pig. He discussed previous contributions and gave his own description of bifurcation cartilages.

I. Brenek (1941) also used this method in an eleven week old foetus to study the state of differentiation of cartilage at that age.

J. Hayward and L. Reid (1952) made an important contribution to the knowledge of normal human bronchial cartilage before they studied the influence it had on bronchiectasis and massive collapse of the lung. In the extra-pulmonary airways they found "C" shaped rings and a membranous part; in the intrapulmonary ones, irregular plates which surround each bronchus completely, but they found no clear-cut transition between the two. The plates were large and closely spaced proximally and tiny, widely dispersed distally. In the proximal part the long axis of each piece of cartilage showed no constant alignment: some had their long axis lying transversely like the rings in the trachea and the main bronchi, in others the long axis was oblique or longitudinal. These authors were the first and only ones who published a drawing of the cartilages surrounding the bifurcation of the left main bronchus in upper and lower lobe bronchi. They also described the bifurcation cartilages more accurately than had been done before.

Finally A. Delmas and I. Eralp (1954) repeated R. Heller and H. von Schroetter's study of the cartilaginous patterns around the tracheal bifurcation and in the carina according to the same criteria and with seemingly the same results.

This historical chapter cannot be concluded without emphasizing that all the authors who recently reviewed the historical evolution of our knowledge of broncho-pulmonary anatomy: A. A. Rap and G. J. Smelt (1947), N. P. D. Smyth (1949), P. Lucien and A. Beau (1951) and E. A. Boyden (1955) took the year 1880 and the publication of "Der Bronchialbaum" as the starting point. Did W. Ewart (1889) indeed not state: "The Literature can be said to consist of a single book (Aeby's)" and A. Narath (1901): "Mit Aeby beginnt eigentlich erst die Geschichte des Bronchialbaumes".

The breach with the past, wanted by Aeby, has thus been perpetuated, but at the same time all interest lost in what was known about cartilages before him. And this may explain why modern text-books are less explicit about them than the ones which appeared at the end of the preceding century.

It should therefore be no matter of surprise that to the question where rings become plates contradictory answers are found in the present literature and that modern science is unable to provide any information on how exactly cartilages are arranged around the bifurcations of the main into the lobar bronchi.

For completeness sake a couple of illustrations should be mentioned which appeared in the 8th (1908) till 19th (1955) edition of A. Rauber and Fr. Kopsch's "Lehrbuch und Atlas der Anatomie des Menschen" and which have been reproduced with some modifications in other text-books: G. Paturet (1958), J. Sobotta and H. Becher (1965).

Together with F. Merkel's (1902) sketch these pictures may be considered as introducing this contribution because they reproduce the facts which are described later. But as they have remained isolated and show some minor inaccuracies, they have contributed little to a better knowledge of tracheo-bronchial cartilages.

III. Method und Material

A. Method

The methods which have so far been used in the study of tracheo-bronchial cartilage are: dissection, histological sections and reconstruction.

Dissection of cartilages was carried out with satisfactory results by R. Heller and von Schroetter (1897) and by A. Delmas and I. Eralp (1954) in the tracheal bifurcation and by J. King (1840) in the smaller bifurcation cartilages. F. Hayward and L. Reid (1952) state they were able to study the shape of the plates in the smallest bronchi, by opening them longitudinally and dissecting off the layers internal to the cartilages. But the number of lungs examined this way was not larger than twelve and in each only four segments were investigated. Which shows that this method does not seem to be appropriate for a comprehensive study of all the cartilages in a large series of bronchial trees.

Histological methods have contributed tremendously to the knowledge of cartilage as a tissue, its structure, chemistry, differentiation, pathology, relationship with the other tissues of the bronchial wall. But they seem to be misleading when it comes to show the shape of the cartilages themselves. This was demonstrated by W. S. Miller (1937): a horizontal section through the trachea, the cartilages of which are obviously rings may suggest the presence of plates when rings are not strictly horizontal.

Reconstructions after Born on the other hand do meet all requirements, but the fact that only two of them are recorded in the literature (W. S. Miller, 1937; I. Brenek, 1941) proves that they are too time consuming to be used routinely.

A review of the three conventional methods of investigating tracheo-bronchial cartilages does therefore show that none is entirely suitable for studying the many intricacies of the cartilaginous skeleton of the bronchial tree in a large series of cases. It does explain why no systematic or comprehensive study has been carried out so far and that no such undertaking could be contemplated unless an entirely different way of investigation was adopted.

1. Macroscopic Staining of Cartilage

Introduced by O. Schulze (1897) and J. van Wyhe (1902), greatly improved and widely used by H. Lundvall (1904, 1905, 1912, 1927), whose name remains attached to it, modified by J. Moreira da Rocha (1917) and C. H. Miller (1921), the mascroscopic staining method has been extensively used in the investigation of skeletal cartilage in fishes, embryos and so on; but to the best of our knowledge, never yet in the study of tracheo-bronchial cartilage.

The basic principle is that after staining whole preparations, the dye is cleared from the soft tissues and the preparation made transparent in an oily mixture, so that only the cartilages which have retained the dye remain conspicuous.

Following variations of the method were used in this study:

a) *Adult human and large animal* bronchial trees, stripped from lung parenchyma were, after thorough fixation in formalin, further fixed for about one week in 70° alcohol. Staining also lasted one week in a solution of:

Methylene blue	2.5 Gm
Hydrochloric acid	30.0 ml
70° alcohol	970.0 ml

Decoloration of other than cartilaginous tissues took place in successive 70° alcohol baths, each lasting one to two days, followed by 95° alcohol during one day. Dehydration took four hours in absolute alcohol and the preparation was finally made transparent and preserved in a mixture of:

Benzyl benzoate	1 part
Methyl salicylate	3 parts.

b) In *small animal and foetal* preparations fixation was also in formalin, but 95° alcohol was used at once for about a week and staining also took one week in

Methylgreen	5 Gm
Acetic acid	330 ml
95° alcohol	660 ml

Decoloration took but one day in 95° alcohol, dehydration four hours in absolute alcohol and the same oily mixture was used for transparency and preservation.

c) For *the study of the hilum* (Chapter IX), whole adult lungs were used. Prior to staining, they were expanded by filling them with formalin through a cannula inserted in the trachea. This reproduced conditions during life and the entrance of the bronchi into the lungs could be precisely demonstrated and photographed (Fig. 29 and following). Too bulky for staining the lungs were reduced in size, by cutting away the distal parenchyma and keeping the lung substance around the hilar structures. After staining with methylene blue, the presence of rings or plates could be noted and after stripping the parenchyma, the stained preparation compared with the pictures of the unstained one and the exact place where the rings become plates, located.

2. Study of the Cartilages

Observation of the cartilages in large preparations had to be carried out in the oil bath against an illuminated back-ground and in small ones in direct light with the aid of binocular magnifying glasses.

In each part of the bronchial tree type, shape, situation and relationship with neighbouring cartilages of each individual cartilage were noted, special attention being paid to connections; rings were counted, all according to criteria to be discussed further.

3. Reproduction

Only small specimens containing little cartilage from young foetusses or very small animals could be photographed as they were and yield clear pictures. In most cases super-position of circumferential cartilage required dividing of the bronchial trees in some or other way to show the shape of the cartilages clearly.

The majority of preparations were cut along a frontal plane, both halves pressed between two glass plates and photographed separately. This was called: "frontal cut" (Figs. 1 to 4). "Lateral cut" means that both main bronchi were separated from the trachea, below its bifurcation, cut laterally and the parts spread between glass plates for taking pictures (Figs. 20, 22 E and 24 E). A "middle lobe cut" was a frontal cut in which the middle lobe bronchus was left attached to the frontal half and cut at the back before being spread out and photographed (Fig. 14).

Of each preparation records were kept including photographs and the characteristics of all cartilages as they were observed on undivided bronchi.

Photographs did show the general appearance of the cartilaginous skeleton well (see next chapter), but details were lost in artefacts and false shadows. In order to show the outline of the individual cartilages unequivocally a number of drawings of selected specimens were made by projecting or enlarging the negatives of the photographs and checking every detail on the original preparation. These drawings were used to illustrate the descriptive part of this monograph.

But however beautiful the pictures, however clear the drawings, observations were not entirely reliable unless made on entire stained preparations in the oil bath, before any cutting or dividing, because it was found that after division back ends of rings very much looked like plates and the relationship between cartilages was recognized less easily.

4. Advantages

Staining of well differentiated cartilages is technically easy and yields a high percentage of excellent results, even if the proposed time-table is not strictly followed. In our adult series for instance, fixation in formalin took between 5 days and 2 months; alcohol treatment 3 to 10 days; staining 3 to 18 days and decoloration 3 to 10 days. Overstaining is desirable, in which case decoloration is likely to last longer and require a greater number of baths. Sticking to the given time-table is much more important in foetal lungs and small specimens.

All manipulations can be carried out at room temperature and several preparations treated simultaneously. Preservation in the oil bath is indefinite and an unlimited number of specimens can be collected. Some have been stained twelve years ago and are still as bright as then.

A last and most important advantage is that the method shows the cartilages sharply outlined and their contour clearly defined, giving a comprehensive picture of every one of the numerous cartilages which go into the skeleton of the whole bronchial tree.

5. Disadvantages

Understaining may be due to technical errors or chemical changes in the cartilages. J. Moreira da Rocha (1917) says that it happens in cartilage which is not purely hyaline and it does occur more often in peripheral bronchi. We have not investigated this histological problem any further and have discarded preparations in which staining was unsatisfactory.

After some time oxidation of the stain may set in and the colour fade. This was not much of a problem in well differentiated cartilage, but in young foetal cartilage it did occur after only a few weeks. It could be delayed by keeping the preparations in a refrigerator in the dark.

During manipulations some distortion and shrinking of the specimens did occur which made them unfit for absolute measurements of lengths of bronchi and angles of bifurcations. Distortion was further increased by dividing and spreading them between glass plates and it may be repeated here that observations were not reliable unless they were made on whole preparations before cutting them.

This disadvantage could be avoided by filling the airways with fusible metal before staining them, but this procedure took too much time and as the specimen ceased to be transparent, it was much more difficult to recognize the exact contour of the cartilages.

In spite of these shortcomings, the macroscopic staining method after Lundvall proved extremely valuable in the investigation of tracheo-bronchial cartilage and it may be recommended without hesitation for further application.

Table 1. Series of 30 foetal lungs in which staining succeeded

No.	Weight (g)	Crown rump length (mm)	Foot length (mm)	Head modulus	Age approximative in days	in weeks
F 1	0.9	21	—	—	50	7
F 2	1.5	24	3.5	—	58	9
F 3	1.7	27	3.7	—	60	9
F 4	2.1	29	4	—	63	9
F 5	3	34	5,5	—	68	10
F 6	4.2	41	6	40	69	10
F 7	4.5	45	—	—	73	11
F 8	6.5	49	7	46	74	11
F 9	8.5	46	—	—	77	11
F 10	9	50	—	—	77	11
F 11	—	53	—	51	80	12
F 12	9.5	57	7.5	—	82	12
F 13	13	61	9	58	84	12
F 14	18	66	10	66	87	13
F 15	19.5	68	10	66	88	13
F 16	22.5	71	10	67	89	13
F 17	25.5	77	11	68	90	13
F 18	26.7	79	11	68	90	13
F 19	65	101	16	92	102	15
F 20	140	115	21	117	112	17
F 21	142	120	24	130	118	17
F 22	207	140	27	134	127	18
F 23	227	147	28	137	131	19
F 24	362	170	34	150	147	21
F 25	374	175	36	157	148	21
F 26	389	180	40	162	148	21
F 27	430	185	38	166	153	22
F 28	488	210	41	162	159	23
F 29	596	200	43	179	165	23
F 30	premature born at 6 months				168	24

B. Material

The parts of the bronchial tree included in this study are the distal end of the trachea, the main and lobar bronchi and the proximal part of the segmental bronchi.

Of 148 normal airways collected on fresh corpses in the autopsy room, 100 were successfully stained and numbered from 1 to 100 at the end of the investigation. Two of these were filled with fusible metal. They were all stripped of lung tissue and used for the description of the cartilages. In 51 a frontal cut was used for reproduction, in 27 a lateral one and in 20 a middle lobe cut; the 2 filled with metal were left undivided. In 47 age and sex were recorded.

A series of 30 whole lungs (7 from large foetusses, 2 from children and 21 from adults) were used for the study of the hilum. They were identified by the letter H and numbered from H1 to H30.

The appearance of cartilage in the human foetus was studied in 30 foetal lungs of increasing age selected from a series of 87. Table 1 shows the parameters which were used for calculating their age, taking into account that the number of days is approximate (G. L. Streeter, 1920).

A number of animal preparations belonging to tetrapods were finally included in this study.

The total number of successfully stained preparations used in this investigation was 173 of which a large number of photographs were taken and the drawings made which illustrate this monograph.

IV. The Adult Human Bronchial Tree

A. General Appearance

As preparations became transparent in the oil bath, an unusual looking, zebra-like, blue coloured cartilaginous skeleton made its appearance. All cartilages were sharply outlined, showing clearly their situation, shape and connections. Each bronchial tree had a similar outlook and yet each had a character of its own. One was supported by slender, regular cartilages, the other by intricate patterns of heavily fused ones.

At first sight rings were seen in the trachea, the main bronchi and the upper and middle lobe ones while the lower lobe bronchi and the segmental ones contained smaller pieces of cartilage. Both main bronchi looked more symmetrical than usually seen in casts and on bronchographies (Fig. 1).

In the anterior half the presence of the same type of cartilages in the main and upper bronchus on the left, the main and middle bronchus on the right, showing some kind of continuity was obvious in all preparations, more outspoken in one than the other, but striking in some (Fig. 2).

In the posterior half the membranous part extending between the back ends of the rings was clearly seen as well as the place where it ceased (Fig. 3).

On close inspection no two preparations were exactly alike. Some characteristics of their cartilages were present each time, others were extremely versatile and where the airways bifurcated, odd combinations were found. It took quite some time before it was possible to decide which features were constant and which were not.

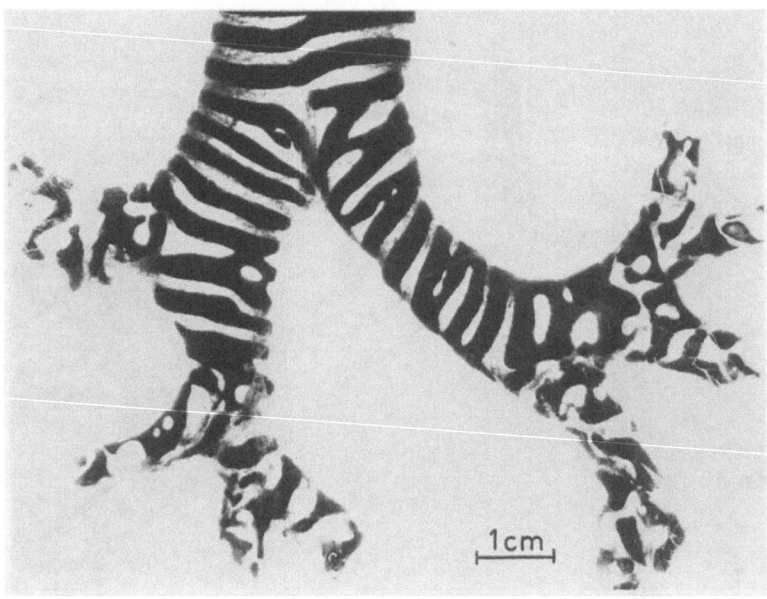

Fig. 1. Photograph showing general appearance. Prep. 5, ♂, aged 61, frontal cut, anterior half

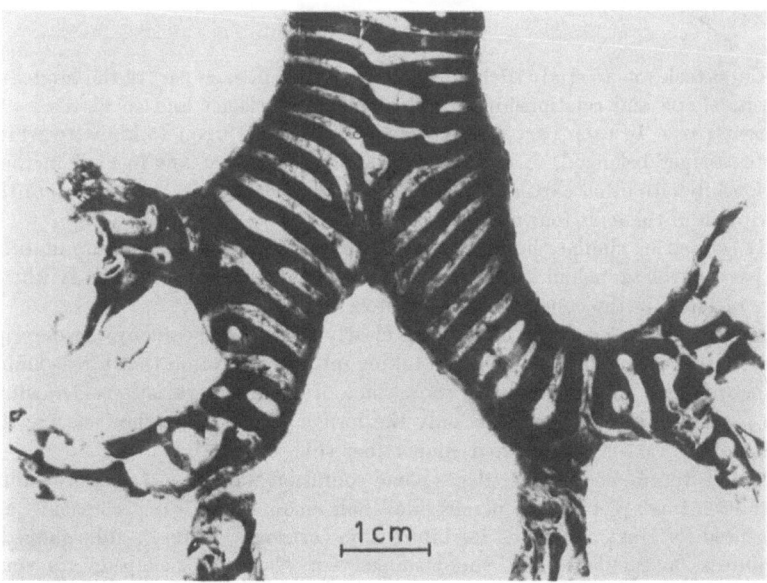

Fig. 2. Prep. 6, ♂, aged 78, frontal cut, anterior half. Symmetry between the two main bronchi and cartilaginous continuity between main and upper and middle lobe bronchi are striking

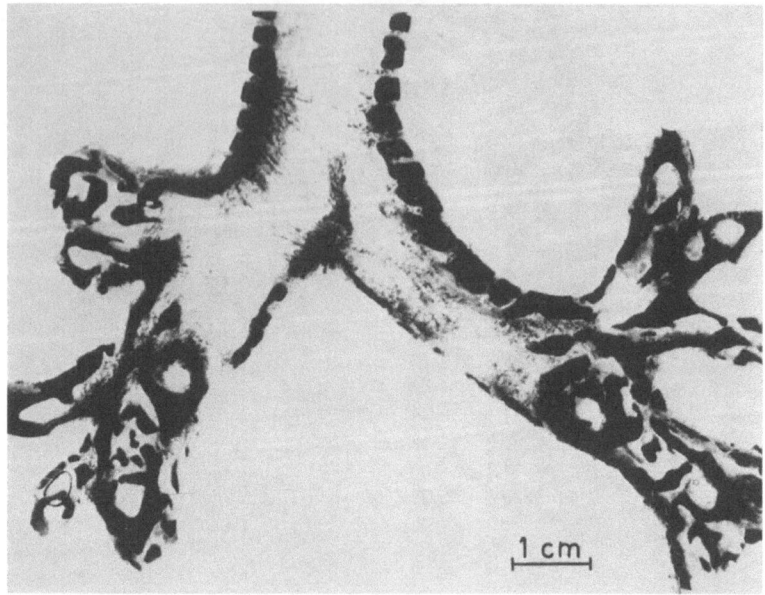

Fig. 3. Same preparation as Fig. 1, posterior half. Shows the extent of the membranous part

A first task was to study each individual cartilage in each part of the bronchial tree, its shape and relationship to neighbouring cartilages and to cartilages in adjacent parts. In each part criteria had to be decided upon to know to which type cartilages belonged, to count how many were present ant to explain their relationship with other cartilages. This analytical study was the foundation of the description in the next four chapters.

By comparing similar characteristics of the cartilaginous architecture in different parts of the bronchial tree, analogies were found which resulted in a synthetic study included in the conclusions to chapters VII and VIII.

Faced with the necessity of showing clearly what was meant by the descriptions, illustrations had to be included taking into consideration that a reasonable number of them were required to show some of the many variations. Drawings were preferred to photographs as only the former were able to show clearly the outline of the cartilages and their connections (Fig. 4).

As to a terminology of cartilages some confusion was found to be prevailing in the literature: picturesque names like "half-moon-shaped" or "ellipsoid" are being used for rings as well as for bifurcation cartilages, while "saddle-shaped" for bifurcation cartilages and "hoop-shaped" or "horse-shoe-shaped" for rings sound little scientific.

Agreement has therefore to be reached upon a terminology which can be consistently used throughout the following pages.

Fig. 4. Drawing of Fig. 1. Illustrates the difference between a photograph and a drawing. (N.B. An all drawings the 1 cm scale is indicated)

B. Terminology

1. Rings

These are the C-shaped cartilages described by Cl. Galenus, which surround about three quarters of the airways front and sideways, the posterior ends or legs of which are joined by the membranous part. Its prototype is the tracheal ring and when similar cartilages are found in other parts of the bronchial tree they may be called "of tracheal type". The term "ring", though inappropriate will however for tradition's sake and because of its historical implications be unashamedly used throughout this monograph, agreeing, that everytime "incomplete rings" are meant.

2. Modified Rings

Where rings become plates, cartilages were often found which were shorter than rings and larger than plates and for which a new name had to be introduced. They might have been called "intermediate" or "transitional" cartilages, but as they behaved in many ways like rings they were called "modified rings".

3. Plates

Their name speaks for itself: they are the small cartilages, extremely versatile in shape, described for the first time by Cl. Bauhinus (1621) in the peripheral airways.

4. Carinal or Bifurcation Cartilages

In the spur, separating bronchi where they divide, the odd-shaped, characteristic cartilages are found which are known as carinal or bifurcation cartilages. By adding the adjectives: primary, secondary, tertiary etc. according to the order of branching to which they belong, no doubt is left as to their localization; a secondary carinal cartilage, for instance, supports a bifurcation of the second order.

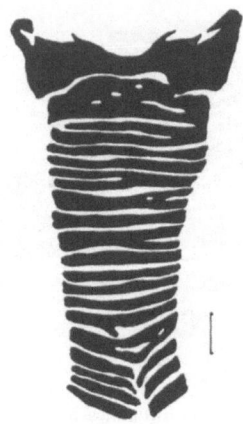

Fig. 5. The stained cartilages in the trachea

V. The Lower Part of the Trachea

Many publications still illustrate tracheal cartilages as single, regular segments or discs neatly piled one above the other as the 16th century anatomists did. This simplified version can only be considered as a diagrammatic representation of the real situation which is shown in the photograph of a whole stained trachea (Fig. 5). This picture does confirm previous statements that tracheal rings may fuse, be broader or smaller, incomplete, forked or X or Y-shaped (G. Paturet, 1957) and agrees with W. S. Miller's (1937) reconstruction.

In our material only the lower end of the trachea was available and in this part the different shapes of the rings could be studied.

A. Polymorphism of Rings

In the great majority of cases rings were free, which means that they were separated from each other by regular, parallel bands of fibrous tissue. Their course was horizontal or slightly sloping, except the last ones which were curved downward in their middle (see next chapter).

Single rings could be shorter than usual or incomplete (not to be confused with intercalar cartilages, see p. 10) fenestrated, broader at one end or L-shaped, curved twice or S-shaped or forked at one end or Y-shaped.

Interrupted rings were sometimes met at the end of the trachea but they were considered as belonging to the main bronchi and not to the trachea (see next chapter).

Compound rings resulted from the fusion of two or three adjacent rings, imitating a number of letters of the alphabet.

The single and compound rings which were encountered in the lower part of the trachea of the adult human bronchial trees are illustrated in Fig. 6.

More complicated compound cartilages in which more than three rings took part were rare. They occurred once along the course of the trachea itself but more frequently in the neighbourhood of the bifurcation (Fig. 9 A).

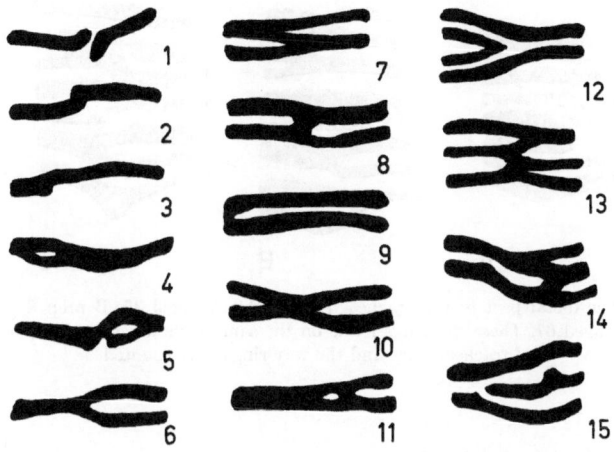

Fig. 6. Single and compound rings found in the lower trachea. *1* interrupted, *2* S-shaped,
3 L-shaped, *4* and *5* fenestrated, *6* Y-shaped or forked, *7* N-shaped, *8* H-shaped, *9* U-shaped
10 X-shaped, *11* A-shaped, *12* V-shaped, *13* and *14* M- or W-shaped, *15* incomplete

The frequency in which rings of different shapes were present could unfortunately not be studied as only part of the trachea was available.

B. Method of Counting Rings

Fig. 7 in which three distal tracheas are shown is intended to illustrate the
difficulties which arose when an attempt was made at counting rings.

Single rings and many compound ones which were obviously the result of
fusion of two or three single rings were easily counted as in Fig. 7A.

But in Fig. 7B things were different. The first partial ring amounted to only
half a ring, but the next forked one to one and a half making in total two rings
according to the principle of "compensation". There was also compensation
between the 4th L-shaped and the 5th half-ring in Fig. 7B.

The second forked ring in Fig. 7B undoubtedly represented only one ring, but
could the same be said of the two lowermost forked rings? Certainly not. Each
of them was either the equivalent of one and a half ring (total three) or of two
rings having one part in common (four rings). The same problem arose in the
two Y-shaped rings in Fig. 7C.

To solve this problem the principle of "space occupation" was introduced.
In Fig. 7B the equivalent of five rings in the upper part occupied a space of
2 cm or an average of 0.4 cm each. The two forked rings occupied a space of
1.5 cm or the space taken by 1.5 cm divided by 0.4 cm = 4 rings approx. The two
forked rings in Fig. 7B were therefore considered as representing four rings. In
Fig. 7C on the other hand the first three rings occupied 1.5 cm, the two forked
ones also 1.5 cm. In that case, the forked rings were considered to correspond to
three rings.

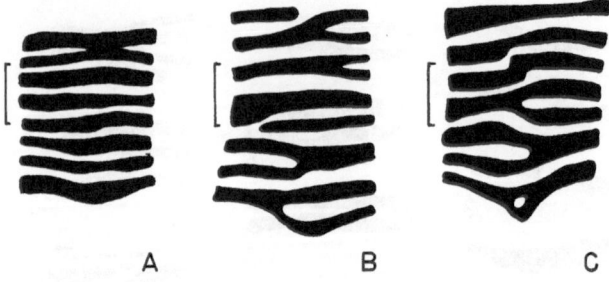

Fig. 7 A—C. The distal part of the trachea. A prep. 11, ♀, aged 35. B prep. 8, ♀, aged 75. C prep. 10, ♀, aged 67. These drawings made on the same scale illustrate the polymorphism of tracheal rings and the way rings were counted

These three examples did illustrate that counting of rings could not be mathematically correct, that one and the same interpretation had to be used in all cases and that only the average of a large series of cases could give reliable figures.

The three distal tracheas in Fig. 7 also illustrate that space-occupation of rings varied in different preparations: although all three were drawn on the same scale 8 rings were counted in A, 7 in B and 6 in C in the first three cm.

The cartilaginous skeletons of tracheas did therefore look different, not only because of the changing shape, width and combination of rings, but also because of the changing distance which separated rings from each-other. Later we shall see that changing patterns were sometimes found in one and the same preparation.

Fig. 7 finally showed that one trachea might be supported mainly by single rings as in Fig. 7 A the other by modified or compound rings as in Fig. 7 B and C.

VI. The Tracheal Bifurcation

In no other part of the bronchial tree have the cartilages been so thoroughly investigated as in the tracheal bifurcation. In the most outstanding contribution to this subject R. Heller and S. von Schroetter (1897) submitted 128 human and 48 animal tracheal bifurcations to dissection and classified them according to the amount of cartilage in the fibrous ridge, "der Theilungsfirst" or carina, separating the two main bronchi. Thirty seven black and white drawings illustrated their publication.

Carinas were considered cartilaginous or membranous depending on the presence or not of cartilage in them. Mixed were the ones in which only part of them was supported by cartilage.

Cartilaginous carinas were called either tracheal, tracheo-bronchial, bronchial left or right, or double bronchial, depending on the origin of the rings which supported them.

They traced this origin by noting whether the corresponding rings ended laterally either above or below the tracheo-bronchial angle, admitting that this angle was obtuse and its apex often hard to define.

A. Delmas and I. Eralp's (1954) publication in which 49 male and 51 female preparations were studied was less explicit and no illustrations included. They divided the cartilaginous structures around the carina in five types: type I corresponded to the tracheal and double bronchial carinas of previous authors, type IV to the membranous and type V to the mixed ones. Types II and III were the ones in which the last tracheal ring was fused either to the first bronchial ring on the left (type II) or on the right (type III). They thus introduced a new concept: fusion between tracheal and bronchial cartilages on one or other side, in other words: cartilaginous continuity between trachea and bronchi.

These two contributions did therefore not seem to use entirely comparable criteria and the confusion which resulted was further increased by giving contradictory figures. In neither paper were main bronchial rings, except the proximal ones studied and the fact of considering the carina to belong to the trachea in some cases (tracheal carinas), to the bronchi in the others (bronchial carinas) even to both (tracheo-bronchial) added to the difficulties of counting rings in the main bronchi as set out in the next chapter.

A. Own Criteria

Clearly defined criteria had therefore to be agreed upon before proposing a new description and classification.

At first, the two different aspects of the problem: cartilaginous continuity between trachea and bronchi on the one hand and cartilaginous support of the carina on the other had to be separated one from the other.

An attempt had then to be made to establish a boundary between trachea and bronchi and the rings which belonged to either part. It soon appeared however that where the trachea divided into two, rings of a same type were found in the three branches making up the bifurcation. Their shape and tendency to fuse were exactly alike and the transition from one part in the other was smooth, undisturbed and did not obey any hard and fast rules. Sometimes there was no difficulty in deciding to which part rings belonged, but often there were two or three possibilities and conventional criteria had to be accepted. The following were the ones which were used in this study:

1. The Tracheo-Bronchial Angle

This landmark which was the only one used by previous authors coincides with the broadening of the trachea and its apex indicates the origin of the main bronchi. Like them we found that that this angle was sometimes obtuse and its apex hard to define. In some cases no angle existed at all.

2. The Carina: A Bronchial Structure

The carina and the cartilaginous rings supporting it was always considered to belong to the main bronchi and never to the trachea. This postulate which agreed with H. von Luschka's description (see p. 10) avoided the complication of classifying carinas in tracheal, bronchial or tracheo-bronchial.

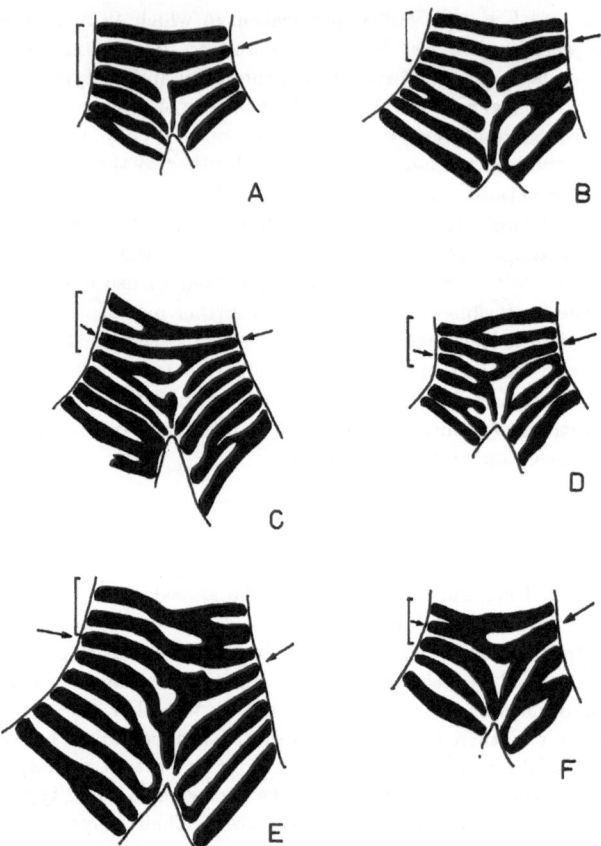

Fig. 8. A prep. 11, ♀, aged 35, last tracheal and first bronchial rings all single. B prep. 60, last tracheal ring lying at a higher level than usual. C prep. 23, ♂, aged 54, cartilaginous continuity between trachea and right main bronchus. D prep. 33, fusion between last tracheal and first bronchial rings on the right. E prep. 6, ♂, aged 78, medial fusion of first bronchial rings and cartilaginous continuity between last tracheal and first bronchial ring. F prep. 66, same relationship of cartilages as in preceding preparation but easier interpretation

N.B. The arrows in this chapter point to the last tracheal ring as defined in the text.

3. The Shape of the Rings

Cartilaginous rings being adapted to the contour of the airways, it was to be expected that where a change in the shape of the air-tubes occurred, a change in the shape of the rings would also take place. In all preparations one ring was found near the bifurcation, showing a downward curve in its middle, unlike the preceding ones which ran straight. This ring was considered to be the last tracheal one. Immediately beyond it two rings, either separate, fused medially or with the last tracheal ring curved upward and were considered to be the first bronchial ones.

By taking the three preceding criteria simultaneously, and not just one of them, into consideration, the last tracheal ring and the first bronchial ones could be

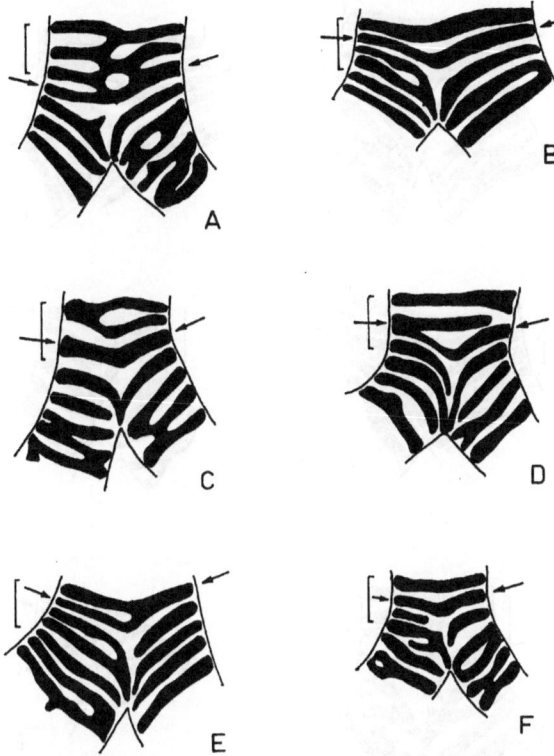

Fig. 9. A prep. 62, last tracheal rings compound, first bronchial ones single, symmetrical distribution of cartilage. B prep. 76, first bronchial ring on the right compound, the others single. C prep. 55, no cartilaginous continuity between trachea and bronchi, first bronchial rings fused medially. D prep. 26, ♀, aged 42, medial fusion between second bronchial ring on the right and the first on the left. E prep. 57, compound rings in trachea and right main bronchus, single in left main bronchus. F prep. 69, unusual symmetrical arrangement of rings

identified in all preparations with a fair degree of consistency, a boundary recognized between trachea and main bronchi and all bifurcations included in a classification in which fusion of rings or cartilaginous continuity was the leading parameter.

B. Description and Classification

1. Absence of Cartilaginous Continuity between Trachea and Bronchi

In 77% of cases there was no connection between tracheal and bronchial rings. But the simplest combination in which the last tracheal as well as the first bronchial rings were single were only rarely met with (Fig. 8A and B). More often one or more of these cartilages were of the combined type (Fig. 9A and B).

The last tracheal ring was sometimes found at a higher level than usual (Figs. 8B, 9E and F) as if both main bronchi were running side to side for some distance before separating. In these cases the interrupted ring was taken to belong to the bronchi.

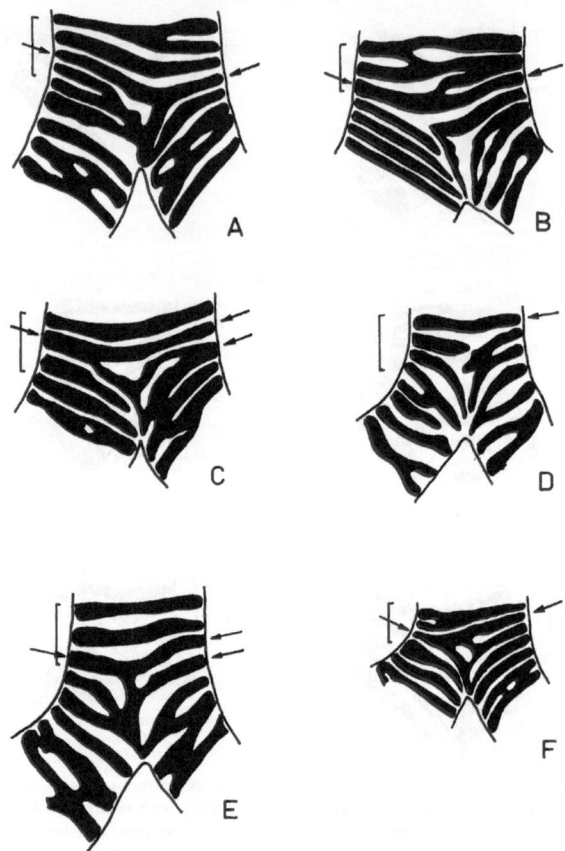

Fig. 10. A prep. 61, intricate pattern with first bronchial cartilages fused medially and the last tracheal to the left bronchial ones. B prep. 58, absence of fusion between tracheal and bronchial rings. C prep. 1, ♂, aged 73, fusion between last tracheal and right main bronchial rings, may also be interpreted as showing left continuity if last tracheal ring is taken one lower. D prep. 70, absence of fusion between tracheal and bronchial rings. E prep. 12, ♀, aged 54, intricate pattern, see text. F prep. 16, ♂, aged 52, intricate pattern, interpreted as bronchial rings fused medially and last tracheal ring fused to first bronchial cartilages, as well on the right as on the left

The first bronchial rings could be fused medially (Fig. 9C) and once the first bronchial rings fused with the second on the other side (Fig. 9D).

2. Cartilaginous Continuity between Trachea and Bronchi

The number of cases in which fusion occurred between trachea and bronchi was smaller. It was found on the right (Fig. 8C and D) as well as on the left side, independent from the fact that fusion was present medially (Fig. 8F) or not.

Table 2. Frequency of fusion between tracheal and bronchial cartilages

right	symmetrical	left
	67 no fusion	
5 right only	10 medial only	7 left only
1 medial and right	2 medial and sides	8 medial and left
Total: 6	79	15

Using the above criteria however did not exclude a small percentage of cases (2 to 3%) in which bifurcations were found the rings of which were so intensely fused that two or three different solutions could be accepted. In Fig. 10C and E, two possible interpretations are shown. The lower arrows were selected because the upward curve of the first bronchial rings was thought to be conspicuous.

These last examples show again that most figures are only approximative and that every attempt at establishing a clear boundary between trachea and bronchi is still conjectural.

3. Recapitulation

The summing up of results in Table 2 showing the percentages of cases in which tracheal and bronchial rings remained separate or fused suggests a predominantly symmetrical distribution of cartilage with variations which may favour the right as well as the left side.

Fig. 11a and b. Frontal section of the carina illustrating. a Cartilaginous; b membranous carinas (after R. Heller and H. von Schroetter, 1897)

C. Amount of Cartilage in the Carina

Previous authors referred to the relationship between cartilage and fibrous tissue in the carinas to classify them as cartilaginous, membranous or mixed (Fig. 11) and left or right bronchial etc. depending on the origin of the rings which supported them.

In our preparations fibrous tissue had shrunk so much during manipulations that all carinas had to be considered as belonging to the cartilaginous type. It was nevertheless possible to assess whether cartilage entering the carina belonged predominantly to the left (Fig. 8A and B), to the right (Fig. 8C and D) or to both main bronchi (Fig. 9B and D). In the latter case, participation of cartilage in the carina was said to be indifferent. Expressed in percentages cartilaginous support of the carina was indifferent in 24%, predominantly left in 29% and predominantly right in 47% of cases.

D. Conclusion

Cartilaginous continuity and cartilaginous support of the carina are but two of the numerous parameters which shape the anatomy of the tracheal bifurcation, others being, for instance: the size, the inclination, the length of the main bronchi and the relationship of the carina to the sagittal middle plane of the trachea. None of these were included in this study.

But the ones concerned with the cartilaginous skeleton both showed that the bifurcation of the trachea was either symmetrical or in favour of one or other side with no clear-cut advantage for one or other side; in other words that, using a large series, the trachea could be considered as dividing into two branches of the same importance.

Study of the cartilaginous skeleton of the tracheal bifurcation also learned that no clear boundary could be drawn between trachea and main bronchi, the cartilages merging unnoticeably from one part in the other.

Since the criteria which were adopted in this study were not quite the same as the ones used by previous authors, comparison of results made little sense and the need will still be felt for a description and classification on which general agreement may be reached.

VII. The Main Bronchi

Both main bronchi are supported by cartilaginous rings which have the same characteristics as tracheal rings. This simple statement is to be found in most anatomical text-books. It means that main bronchial rings are C-shaped and completed at the back by a membranous part, that they may remain single or fuse and produce compound rings, similar to the ones in the trachea. It means, in short, that main bronchial and tracheal rings are alike.

Little mention is made nowadays of the number of rings which go into each main bronchus. But in text-books dating from the end of last century, figures were published which were fairly constant of the left main bronchus, but variable and generally much lower for the right one (Table 3). The reason was that some anatomists believed that the right main bronchus ended at the level of the right upper lobe bronchus, others that it reached as far as the bifurcation of the middle lobe bronchus.

Table 3. Number of rings in main bronchi. Figures before 1900

	Right	Left
J. F. Meckel (1820)	8	9–12
Ph. C. Sappey (1874)	4–5	9–13
R. Hartmann (1881)	6–8	9–13
Ch. Morel and M. Duval (1883)	±6	10–12
K. Gegenbauer (1889)	4–8	8–12
J. Quain (1896)	6–8	9–12

Our observations on stained preparations entirely confirmed the tracheal morphology of main bronchial rings, but they were found to fuse more easily and produce intricate patterns, defying all description, especially distally. A number of new observations could also be made.

A. The Right Main Bronchus

The problem which had been encountered when attempting to identify the first bronchial ring at the tracheal bifurcation arose again when it became necessary to decide which was the last one belonging to the right main bronchus.

In many cases rings were indeed seen to merge with the middle lobe cartilages (Fig. 13) and when they did not, it was not always easy to decide which rings belonged to the main and which to the lobar bronchus (Fig. 14).

As in the tracheal bifurcation conventional criteria had therefore to be chosen and these were:

a) the place where the main bronchus, due to the fact of bifurcating in middle and lower lobe bronchi, became broader;

b) the apex of the angle between main and middle lobe bronchi;

c) the orifice of the segmental apical lower lobe bronchus which undoubtedly belongs to the lower lobe bronchus.

1. Number of Rings

Using the principles of "compensation" and "space occupation" introduced in the study of the lower part of the trachea, rings could now be counted with

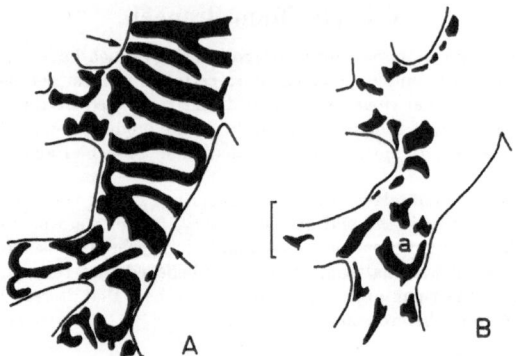

Fig. 12 A and B. Prep. 8, ♀, aged 75, frontal cut. A Anterior half, total 7 rings, 3 predominantly single in upper part, 4 predominantly compound in lower part. B Posterior half, membranous part ends where cartilages supporting the orifice of the apical lower segmental bronchus appear

N. B. In this chapter the arrows point to the first and the last rings; a, orifice of the apical lower segmental bronchus

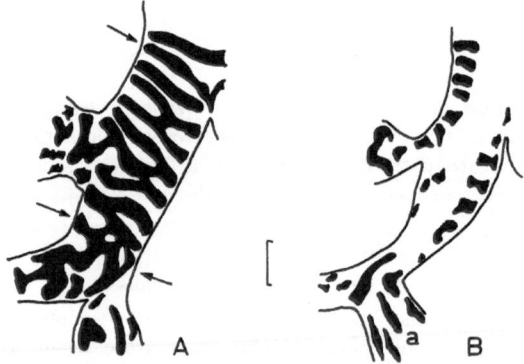

Fig. 13 A and B. Prep. 34, ♀, aged 77, frontal cut. A Anterior half, total 10 predominantly compound rings, 6 in upper, 4 in lower part, connection between upper and lower parts. B Posterior half, apical lower segmental bronchus cut lengthwise

a fair degree of accuracy, admitting that at the beginning as well as at the end of the main bronchus different interpretations were likely to influence results.

With some experience counting became easier and more reliable.

Amongst results recorded in Table 4 emphasis should be laid on the observations that in the highest percentage of cases (34%) 10 rings were found, that in 85% between 9 and 11 were counted and that the overall average was a little less than 10 (9.90).

2. The Upper and the Lower Part

The emergence of the right upper lobe bronchus divides the right main bronchus into two parts: an upper and a lower one, the latter being known as

Table 4. Number of rings in right main bronchus

Number of rings	Number of bronchi
7	4
8	4
9	28 ⎫
10	34 ⎬ 85
11	23 ⎭
12	6
13	1

Table 5. Number of rings in upper and lower parts

Number and rings	Number of bronchi
Upper part	
3	4
4	11
5	29
6	36
7	19
8	0
9	1
Lower part	
2	2
3	9
4	53
5	28
6	8

the bronchus intermedius, a term which was not officially recognized by the P.N.A.

The last ring of the upper part was seen in all preparations to support the upper lip of the right upper lobe bronchial orifice and the first ring of the lower part its lower lip, as if both these main bronchial rings embraced this orifice.

Rings were counted in both parts and found to average 5.61 in the upper part and 4.31 in the lower part (Table 5).

3. Connections

The relationship between main bronchial rings and cartilages belonging to adjacent parts (the trachea, the right upper lobe, the middle lobe and the lower lobe bronchi) were studied in the chapters dealing with these parts.

But all along its course rings in the right main bronchus also showed a tendency to fuse. An appraisal of this tendency was made by taking upper and lower parts separately and considering cartilages "predominantly single" when more than half were single, or "predominantly compound" when more than half were fused in each part. This gave a rough estimate of 30% predominantly

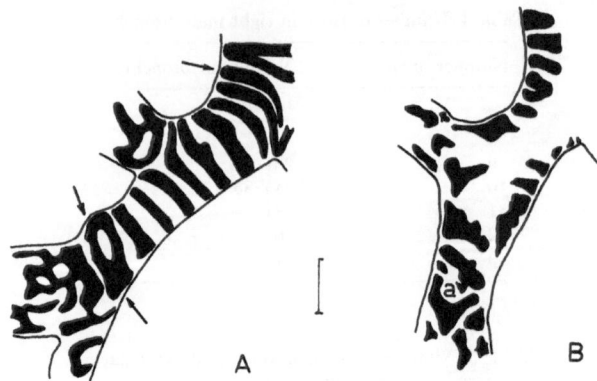

Fig. 14A and B. Prep. 55, middle lobe cut. A anterior half, eleven rings: 6 in upper and 5 in lower part, it is not easy to decide where the main bronchus ends and the middle lobe one begins. B Posterior half, one large plate appears in the distal membranous part

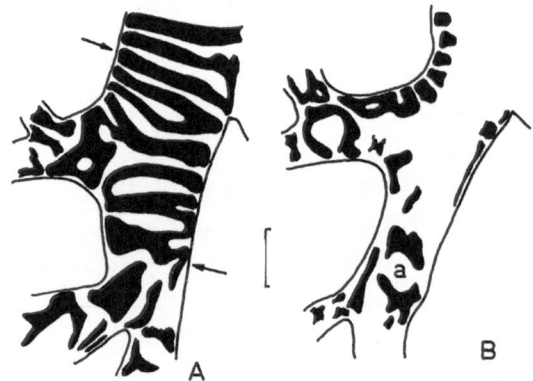

Fig. 15A and B. Prep. 40, ♀, aged 70, frontal cut. A Anterior half, seven rings predominantly compound, 4 in the upper and 3 in the lower part. B Posterior half

single in the upper part and 0% in the lower part, implying an increased tendency of cartilages to interconnect as the main bronchus grew longer.

Fusion between upper and lower part occurred in 33% of cases (Fig. 13) which was also higher than fusion between tracheal and main bronchial rings on this side, 6%.

4. The Membranous Part

A paries membranacea or membranous part, continuous with that of the trachea was present in all right main bronchi. It reached as far as the cartilages supporting the apical lower segmental bronchial orifice and was free from cartilages in 41% of cases. In the remainder one (47%), two (10%) or three (2%) small plates appeared distally as if this membranous part did not end abruptly, but gradually.

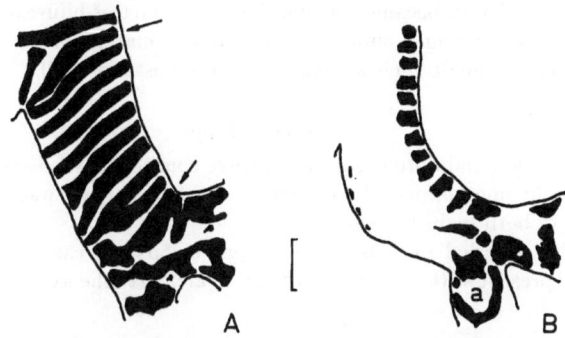

Fig. 16A and B. Prep. 39, frontal cut. A Anterior half, ten rings, predominantly compound in lower half. B Posterior half, membranous part showing two small plates in its distal part

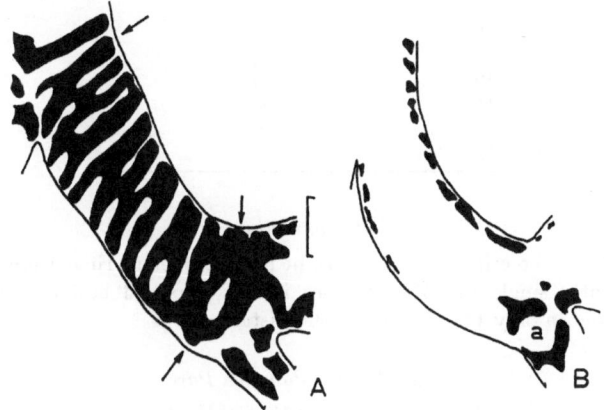

Fig. 17A and B. Prep. 31, ♂, aged 56, frontal cut. A Anterior half. eleven predominantly compound rings. B Posterior half

B. The Left Main Bronchus

Rings of tracheal type were also found to support the left main bronchus. They were either single or fused to each-other, sometimes in a very versatile and compact way, especially where the bronchus ended.

Where the left main bronchus bifurcates in upper and lower lobe bronchi, rings ended abruptly in rare cases only (Fig. 16). Mostly rings of tracheal type were also present in the left upper lobe bronchus to which the main bronchial ones were either connected (Fig. 17) or not.

A precise boundary between the left main and upper lobe bronchi was therefore hard to define and conventional criteria had again to be accepted to recognize which cartilages belonged to either bronchus.

The left main bronchus was conventionally considered to end in the same way as the right one:

a) Where the airway became broader due to the fact of bifurcating.

b) Where the cartilages supporting the upper margin of the apical lower segmental bronchus made their appearance in the posterior wall.

1. Number of Rings

Rings in the left main bronchus were always counted at the same time as the ones in the right one because the cartilaginous physiognomy was usually found to be similar in both main bronchi.

The results recorded in Table 6 show that in 85% of cases between 9 and 11 rings were present with a peak of 37% at 10 and that the average was near 10 (10.20)

Table 6. Number of rings in left main bronchus

Number of rings	Number of bronchi
7	1
8	3
9	21 ⎫
10	37 ⎬ 85
11	27 ⎭
12	8
13	3

2. Connections

Using the same criteria as in the right main bronchus, rings were found to be predominantly single in 20% of cases in the proximal half of the left main bronchus and in only 1% of cases in the distal half.

3. The Membranous Part

A membranous part, continuous with that of the trachea reached as far as the appearance of the apical lower segmental orificial cartilages in the left main bronchus. In 14% of cases no cartilage was found in it. In 42% of cases one, in 31% two, in 4% three, in 8% four and in 1% five small plates were present in the distal part.

C. Comparison between the Two Main Bronchi

Difference in size, length, direction and rotation all contribute to make both main bronchi look asymmetrical in unstained preparations.

But when their cartilaginous framework was investigated, a great number of characteristics were found to be similar.

1. Both contained the same type of rings as the trachea.

2. On either side, between 9 and 11 rings were present in 85% of cases, the highest percentage being 10 and the average very near 10.

3. On the right as well as on the left rings fused more intensely as the bronchi grew longer.

4. Main bronchial rings were continuous with lobar cartilages on either side in a number of cases.

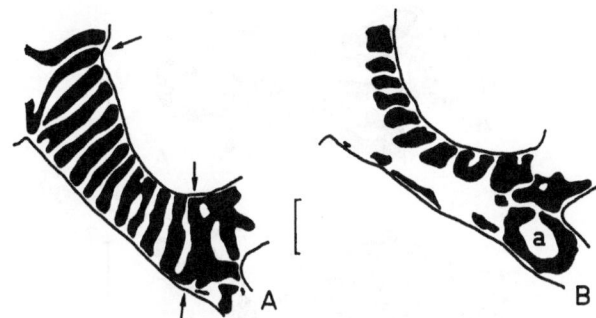

Fig. 18A and B. Prep. 55, frontal cut. A Anterior half, twelve rings. B Posterior half, membranous part with small plate

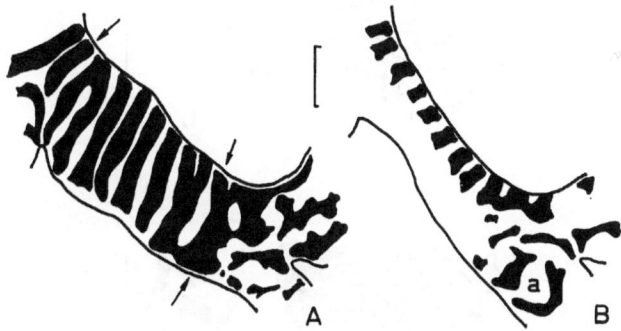

Fig. 19A and B. Prep. 40, ♀, aged 70, frontal cut. A Anterior half, nine rings. B Posterior half, two plates in membranous part

5. In both a similar membranous part ended in the same way.

6. In 37% of cases both main bronchi contained the same number of rings, in 41% a greater number were found in the left main bronchus and in 22% a greater number in the right one.

Taking into consideration its most intimate structure, the adult bronchial tree showed, when stained, two main bronchi which were fundamentally symmetrical in the same way as the trachea was found to divide into two equal branches.

VIII. The Lobar Bronchi

Investigation of the cartilages supporting the lobar bronchi and their junction to the main bronchi was like exploring new territory.

Except a single reconstruction of the cartilages in the right upper lobe bronchus by W. S. Miller (1937) and a single drawing of the cartilages surrounding the bifurcation of the left main bronchus in lobar branches by J. Hayward and L. Reid (1952), no contribution to this subject has been produced as far as we know.

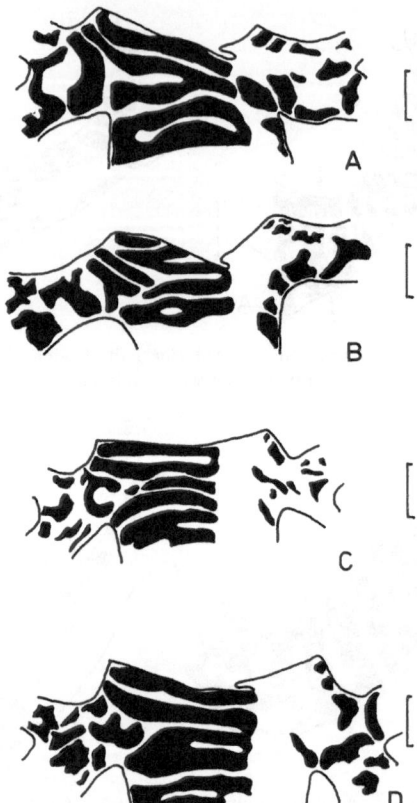

Fig. 20 A—D. Cartilages in the right upper lobe bronchus and its junction to the main bronchus. A prep. 97, lateral cut, one typical ring lying free in the anterior wall, one small plate
lying free in membranous part, other cartilages being back ends of rings and tertiary carinal
cartilages. B prep. 76, lateral cut, three rings in anterior wall, no plates in membranous
part. C prep. 84, lateral cut, crescent-shaped plate in anterior wall, four small plates in posterior
wall. D prep. 94, lateral cut, plates only in anterior wall, no plates in membranous part

N. B. In this chapter the arrows point to the anterior or posterior leg of the secondary
     carinal cartilage; a, orifice of the apical lower segmental bronchus; c, cardiac bronchus

Instead a number of vague and contradictory statements are found in the literature.

1. Large bronchi contain rings, smaller ones plates (most present-day text-books).

2. All lobar bronchi are surrounded by plates (M. Levy, 1951; M. Bariety *et al.*, 1951;
J. Hayward and L. Reid, 1952; D. Kassay, 1961 etc.).

3. All lobar bronchi are surrounded by rings (A. Policard, 1955; G. Cordier and C. Cabrol,
1959).

4. "Large" or main and lower lobe bronchi are surrounded by rings, "medium" or the
middle and both upper lobe bronchi contain plates (H. von Hayek, 1960; V. S. Krahl, 1964).

5. The proximal part of the lobar bronchi are surrounded by rings, the distal part by
plates (K. Dietzel, 1964).

The following description may help in deciding which of these corresponds to
reality.

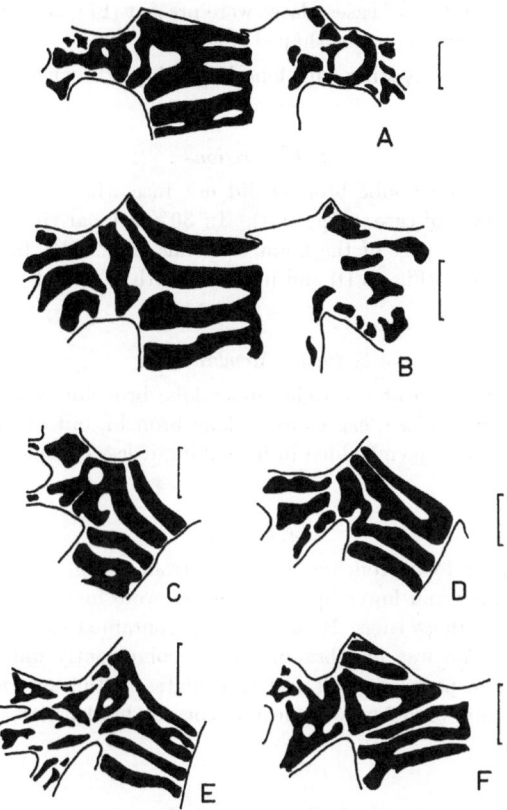

Fig. 21 A—F. Other variations of cartilages in the right upper lobe bronchus and its junction to the main bronchus. A prep. 100, lateral cut, one single ring connected to the last one in the upper part of the main bronchus and one compound ring in the anterior wall, in the posterior wall are the cartilages which surround the orifice to the posterior segmental bronchus. B prep. 86, lateral cut, one ring lying free in anterior wall. C prep. 54, middle lobe cut, combination of rings counted as two, connected to the upper part of the main bronchus. D prep. 57, middle lobe cut, one ring connected to first ring in lower part of main bronchus. E prep. 50, ♀, aged 48, frontal cut, plates in anterior wall of lobar bronchus, one modified ring attached to upper part of main bronchus. F prep. 74, lateral cut, the proximal ring in the lobar bronchus is connected to both upper and lower parts of the main bronchus

A. The Right Upper Lobe Bronchus

1. Cartilages

This bronchus was usually supported by a few cartilages before the tertiary bifurcation cartilages appeared supporting the spurs between its segmental branches.

Rings of tracheal type were found in 73% of cases (Fig. 20A): one in 50% (Fig. 21B), two in 21% (Fig. 20A), three in 2% (Fig. 20B). They were sometimes shorter and belonged to the type described as modified rings (Fig. 21A).

In the remaining 27% of cases plates were present (Fig. 20 D) and one of them was sometimes typically crescent-shaped (Fig. 20 C).

Rings usually went together with long bronchi, plates with short ones.

2. Connections

Plates in right upper lobe bronchi did not fuse with main bronchial rings, but rings did in 33% of cases (Fig. 21 A). In 30% a lobar ring was fused to the last ring in the upper part of the main bronchus (Fig. 21 C), in 2% to the first one on the lower part (Fig. 21 D) and in 1% to both (Fig. 21 F).

3. Paries Membranacea

In the posterior wall of the right upper lobe bronchus a membranous part could usually be recognized, especially in long bronchi, but in 72% of cases one to four small plates were embedded in it, making it less conspicuous.

4. Orifice

The right upper lobe bronchus, emerging from the main bronchus between two rings, the upper and lower lips of its orifice were found to be supported by the lateral ends of these rings. Its anterior lip contained either a ring or plates and its posterior one was membranous. Only coincidently did the presence of small plates in the posterior wall and true plates in the anterior one give the impression that this bronchus was entirely surrounded by plates.

B. The Left Upper Lobe Bronchus

1. Cartilages

Left upper lobe and left main bronchial cartilages were usually found to merge unnoticeably one into the other because rings of tracheal type also supported the lobar bronchus in 97% of cases: one ring in 70% (Fig. 22 E), two in 30%, three in 4%. Only in 3% of cases were plates only found in left upper lobe bronchi.

Consequently a clear boundary between main and lobar bronchus could not be recognized in many cases and had to be based on conventional criteria (p. 33).

2. Connections

The first lobar ring did not fuse to the main bronchial ones in 25% of cases (Fig. 22 E), although lobar cartilages showed a tendency to interconnect between themselves and produce compound patterns.

The first lobar ring was found to fuse with the last main bronchial one in 46% of cases (Fig. 23 A), with the soon to be described secondary carinal cartilage in 12% (Fig. 22 A) and with both in 17% of cases (Fig. 22 C).

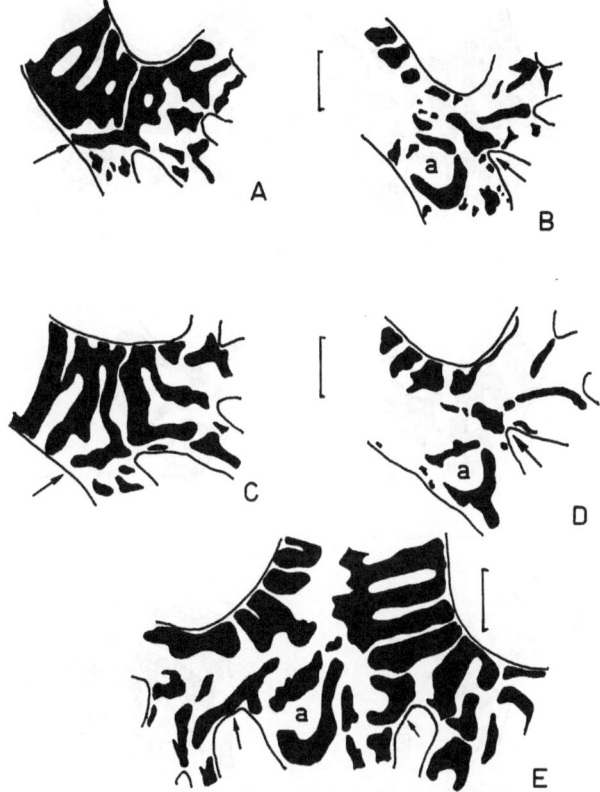

Fig. 22 A—E. Cartilages in the left upper lobe bronchus and its junction to the main bronchus. A prep. 55, ♀, aged 66, frontal cut, anterior half, the lobar bronchus contains two rings, fused together and to the secondary carinal cartilage, there is no fusion between main bronchial and adjacent cartilages. B posterior half, the posterior leg of the secondary carinal cartilage is seen lying between the back ends of the main bronchial and lobar rings and the cartilages belonging to the orifice of the apical lower segmental bronchus. C prep. 7, ♂, aged 35, frontal cut, anterior half, two rings in the lobar bronchus, the first one fused to both the last main bronchial and secondary carinal cartilages. D posterior half, same comment as for Fig. 22 B. E prep. 86, lateral cut, lobar, main bronchial and secondary cartilages are all independent from each-other, the latter is cut into two halves

3. The Secondary Carinal Cartilage

In the spur dividing the secondary bifurcation, in other words, the division of the left main bronchus into its lobar branches, a bifurcation type of cartilage was found for the first time. It could be identified in all preparations. Because of its constancy and the importance it will gain in future deductions, it was felt that it deserved a more elaborate description and a proper name.

a) The secondary carinal cartilage always had the shape of a horse-shoe, the legs of which were pointing in the direction of the main bronchial axis. The length

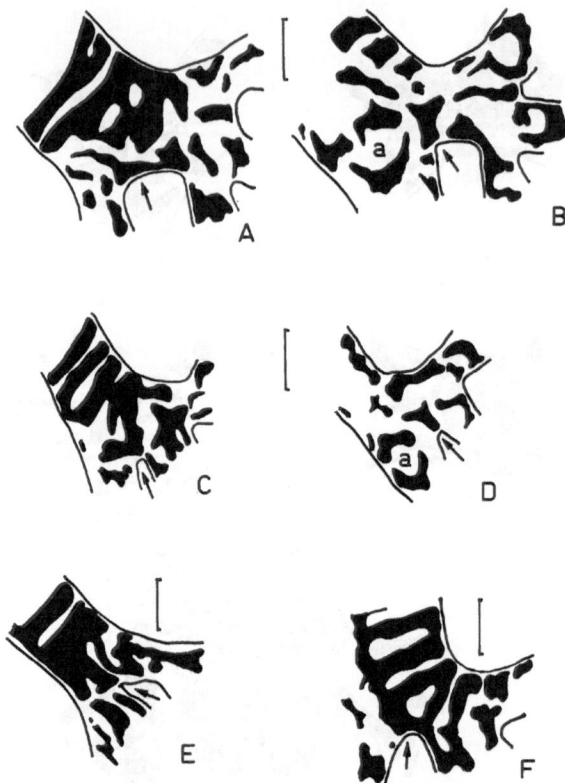

Fig. 23 A—F. Other variations of cartilages in the left upper lobe bronchus and its junction to the main bronchus. A prep. 4, ♂, aged 56, frontal cut, anterior half. B posterior half, note posterior leg of secondary carinal cartilage. C prep. 9, ♂, aged 42, frontal cut, anterior half, one ring belonging to the lobar bronchus fused to the main bronchial and secondary carinal cartilages. D posterior half, the posterior leg of the carinal cartilage and the cartilages supporting the orifice of the apical lower segmental bronchus are always easily identified. E prep. 52, middle lobe cut, anterior half, odd-shaped pattern, complete fusion of last main bronchial ring is rare. F prep. 83, lateral cut, anterior half, the last main bronchial ring, the first lobar one and the anterior leg of the secondary carinal cartilage are all interconnected

these legs had was very variable and could be either very short or reach in front as far as the medial side of the lower lobe bronchus and look like a ring (Fig. 22 A).

b) The anterior leg remained free in 53% of cases (Fig. 23 A) or fused with the main bronchial rings (16%, Fig. 23 E), lobar rings (18%, Fig. 22 A) or both (13%, Fig. 23 F).

c) The posterior leg was always free, lying between the back ends of the main bronchial and lobar rings and the cartilages supporting the orifice to the apical lower segmental bronchus.

d) The central part was sometimes broader and extended into the lateral wall of the lower lobe bronchus.

Unfortunately secondary carinal cartilages did not show very well on reproductions of cut preparations, but they could be easily and unmistakenly identified in undivided, stained bronchial trees.

4. Paries Membranaceus

The posterior wall of the left upper lobe bronchus was seen to be filled by the back ends of the main bronchial and lobar rings, the posterior leg of the secondary carinal cartilage and the bifurcation cartilages supporting the tertiary carinas. A membranous part could therefore not be recognized in left upper lobe bronchi.

5. Orifice

Seen from the inside, the orifice of the left upper lobe bronchus is sharply demarcated only in its lower part by the mucosal fold covering the secondary carina. The reason why its upper part does not show a ridge became obvious in stained bronchi in which its cartilaginous skeleton was seen to be continuous with the main bronchus. The cartilages which participated in the support of the orifice to the left upper lobe bronchus were: part of the secondary carinal cartilage below, the main bronchial and lobar rings above, in front and at the back.

C. The Middle Lobe Bronchus

1. Cartilages

Right middle lobe cartilages, adapted to the smaller size of this bronchus had queer shapes. Many were fenestrated large plates, with bars protruding like pseudopodia. They only rarely looked like rings, but they were possibly the fusion of short rings and as they behaved like rings by connecting to neighbouring cartilages, it was found that the qualification of "modified rings" suited them well.

2. Connections

The cartilages in the right middle lobe bronchus remained separated from the neighbouring ones in 41% of cases (Fig. 24A). In 38% they were fused to the main bronchial ones (Fig. 24C), in 13% to the secondary carinal cartilage (Fig. 24F) and in 8% to both (Fig. 24E).

3. The Secondary Carinal Cartilage

In the spur or carina, marking the secondary bifurcation or division of right main middle and lower lobe bronchi, a bifurcation type of cartilage was also identified in all cases. It was found to be in all respects similar to the one on the left side.

a) It was always horse-shoe shaped. Its legs pointing towards the main bronchial axis had varying lengths in front as well as at the back.

b) The anterior leg either remained free (50%, Fig. 25A) or fused with the main bronchial rings (26%, Fig. 25C), the middle lobe ones (14%, Fig. 25E) or both (10%, Fig. 25F).

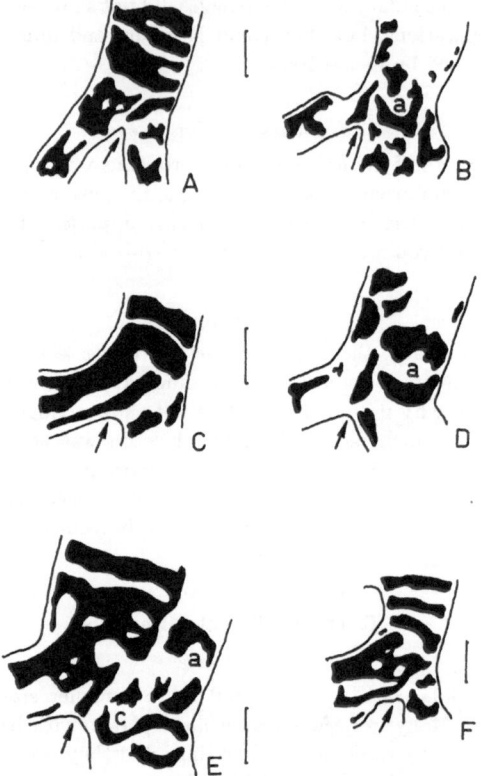

Fig. 24 A—F. Cartilages in the middle lobe bronchus and its junction to the main bronchus, A prep. 4, ♂, aged 56, frontal cut, anterior half, no connection between main bronchial, lobar and secondary carinal cartilages, middle lobe bronchial cartilage combination of two short modified rings. B posterior half, posterior leg of secondary carinal cartilage next to orifice of apical lower segmental bronchus. C prep. 2, ♂, aged 54, frontal cut, anterior half, lobar cartilage connected to main bronchial one, secondary carinal cartilage free. D posterior half, posterior leg of secondary carinal cartilage and cartilages in orifice of apical lower segmental bronchus easily identified. E prep. 89, lateral cut, anterior half, lobar cartilage fused to both the main bronchial and secondary carinal cartilages. F prep. 68, middle lobe cut, lobar and secondary carinal cartilages connected to each-other

    c) The posterior leg always remained free lying between the back ends of the main bronchial and middle lobe rings and the cartilages supporting the orifice to the apical lower segmental bronchus.

    d) Its medial part was either thin or broad, but never to such an extent that it penetrated into the wall of the lower lobe bronchus.

4. Paries Membranaceus

    The posterior wall of the middle lobe bronchus was less heavily supported by cartilage, but because it was filled by the back ends of the carinal and lobar cartilages, no membranous part could be recognized in it.

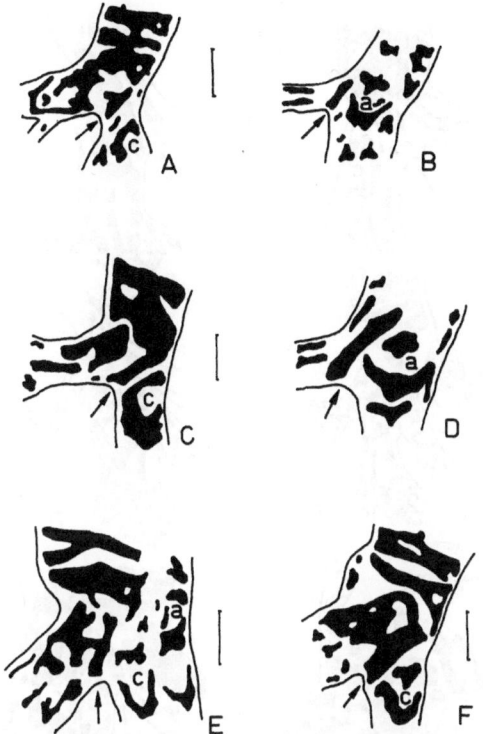

Fig. 25A—F. Other variations of cartilages in the right middle lobe bronchus and its junction to the main bronchus. A prep. 28, ♀, aged 71, frontal cut, anterior half, middle lobe cartilage fused to main bronchial one. B posterior half, usual arrangement of cartilages. C prep. 25, frontal cut, anterior half, middle lobe cartilage free, secondary carinal cartilage fused to main bronchial rings. D posterior half, very long posterior leg of carinal cartilage. E prep. 82, lateral cut, lobar cartilage fused to carinal one. F prep. 67, middle lobe cut, anterior half, lobar cartilage, last main bronchial and secondary carinal cartilages fused

5. Orifice

Seen at bronchoscopy, the middle lobe bronchial orifice has, unlike the left upper lobe one, a sharp mucosal ridge all around it. But on stained bronchi, its cartilaginous skeleton undoubtedly looked like the continuation of the main bronchial one and its orifice contained below the middle part of the secondary carinal cartilage, above and in front the middle lobe cartilages and behind the posterior leg of the carinal cartilage.

D. The Right Lower Lobe Bronchus

1. Cartilages

Unlike the lobar bronchi described so far, right lower lobe bronchi never had any rings in their walls, but plates and tertiary carinal cartilages, evenly distributed around them. The cartilages surrounding the orifices to the apical lower and cardiac segmental bronchus were always conspicuous.

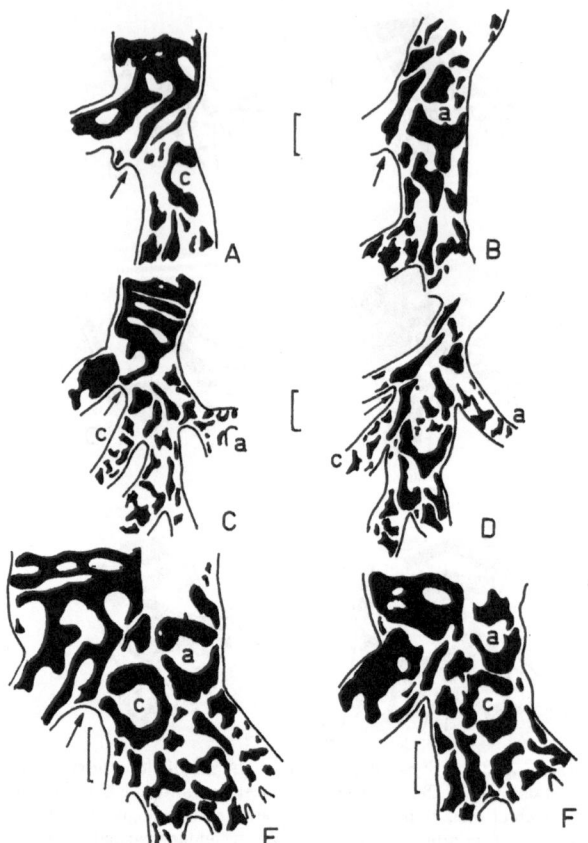

Fig. 26 A—F. Cartilages in the right lower lobe bronchus and its junction to the main bronchus. A prep. 13, ♂, aged 42, frontal cut, anterior half, note tertiary carinal cartilage to cardiac bronchus. B posterior half, presence of plates and tertiary carinal cartilages as in anterior half. C prep. 16, ♂, aged 52, frontal cut, anterior half, in this preparation the apical lower and the cardiac bronchus have been cut lengthwise. D posterior half, a tertiary carinal cartilage in the orifice to one of the basal bronchi is seen. E prep. 81, lateral cut, the lower lobe bronchus being spread out, tertiary carinal cartilages are easily identified. F prep. 75, lateral cut, same comment as for Fig. 26 E

2. Connections

Cartilages in lower lobe bronchi did not show any tendency to fuse or interconnect. Maybe long plates were the combination of two cartilages, but this was the exception and not the rule.

Never did any of the right lower lobe cartilages fuse to the main or middle lobe bronchial ones or to the secondary carinal cartilage. No exception to this rule was encountered.

3. Orifice

Looked at from the inside as in bronchoscopy, the main and lower lobe bronchial mucosal lining run smoothly one into the other and no mucosal fold or irregularity show the boundary between the two, so that the latter appears as the continuation of the former.

The arrangement of the cartilages on the contrary revealed the reverse: the main bronchial rings were seen to be continuous with the middle lobe ones and the lower lobe ones were separated from them by a fibrous band running below the anterior ends of the main bronchial rings, the secondary carinal cartilage and above the orificial cartilages belonging to the apical lower bronchus. This fibrous junction, which was in fact the boundary between the main and lower lobe bronchi was always seen to be lying in an oblique plane.

4. The Orifice to the Apical Lower Segmental Bronchus

The way preparations were cut made the cartilages surrounding this orifice stand out as a clear landmark. They corresponded to the end of the main bronchial membranous part. Usually (69% of cases) one carinal cartilage was present below and a smaller one above, simulating a û. In 16% these cartilages were ü-shaped, in 1% C-shaped and in 2% there was one tertiary carinal cartilage below and three smaller plates above. In 12% of cases their shape was not recorded.

E. The Left Lower Lobe Bronchus

1. Cartilages

Type, shape, situation, size and distribution of cartilages were the same in the left as in the right lower lobe bronchus. Lacking a cardiac bronchus and being slightly longer before deviding in basal branches than the right lower lobe bronchus, more plates went into its cartilaginous support. A long, cartilaginous bar, lying obliquely at its origin and simulating a ring was seen rather frequently in frontal cuts. But in lateral cuts it was obvious that no membranous part corresponded to them that they did not fuse with neighbouring cartilages and that they had to be considered as plates.

2. Connections

Cartilaginous continuity between lower lobe and main or upper lobe cartilages was never observed. A broad secondary carinal cartilage was sometimes seen to penetrate the lateral wall of the lower lobe bronchus for some distance, but it never fused with any of its cartilages. Whether large, complicated cartilages (Fig. 27H) resulted from the fusion of two or several plates remain open to question. They were rare indeed.

3. Orifice

Like the right lower bronchus, the left one appeared to be joined to the main one by a fibrous attachment running obliquely below the anterior ends of the main bronchial and left upper lobe cartilages, the secondary carinal cartilage and above the cartilages in the orifice of the apical lower segmental bronchus.

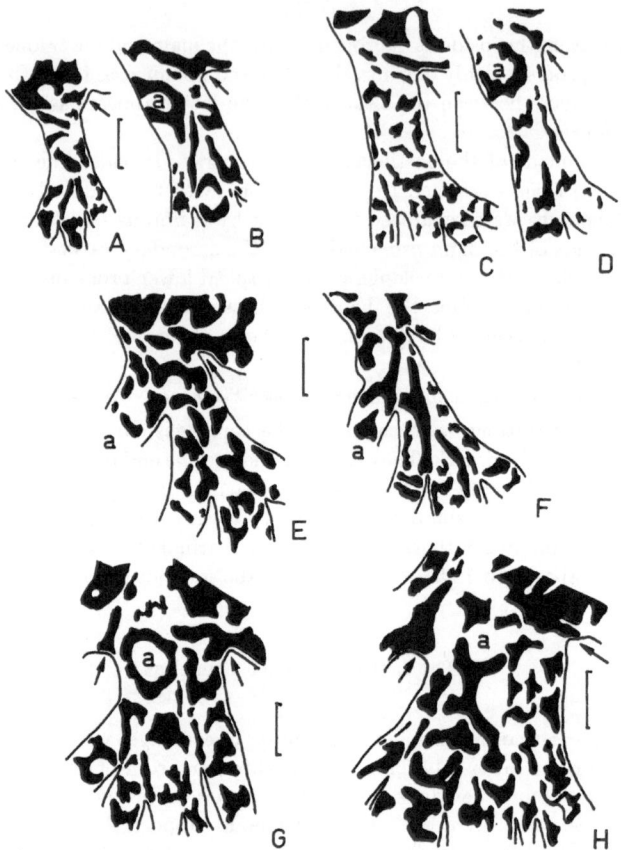

Fig. 27 A—H. Cartilages in the left lower lobe bronchus and its junction to the main bronchus. A prep. 71, middle lobe cut, anterior half. B posterior half, plates and tertiary carinal cartilages evenly distributed around lobar bronchus. C prep. 11, ♀, aged 35, frontal cut, anterior half. D Posterior half, in the left lower as well as in the right lower lobe bronchus fusion between their cartilages and the ones belonging to adjacent bronchi is never observed. E prep. 29, ♂, aged 74, frontal cut, anterior half. F posterior half, apical lower segmental bronchus cut lengthwise. G prep. 77, lateral cut, lower lobe bronchus spread cut shows rare circular cartilage around orifice to apical lower bronchus. H prep. 93, lateral out, the tertiary carinal cartilage to the apical lower segmental bronchus shows long extension which may be plate fused to it

4. The Orifice to the Apical Lower Segmental Bronchus

The cartilages surrounding this orifice were û-shaped in 50%, ü-shaped in 23%, C-shaped in 12%, ̈u-shaped in 2% and O-shaped in 4% of cases. In 9% of cases their shape was not recorded. They indicated the posterior upper limit of the lower lobe bronchus and the termination of the main bronchial membranous part.

F. The Five Lobar Bronchi

The preceding description made it clear that none of the existing theories on the cartilaginous framework of the lobar bronchi and their junction to the main bronchi mentioned at the beginning of this chapter were confirmed by the present research.

The circular cartilages described by F. D. Reisseisen (1822) and "another type of cartilages, circular in shape, which surround a bronchus where it leaves its main-stem bronchus (W. S. Miller, 1937)" were only exceptionally found (Fig. 27G).

An important finding on the other hand was the presence of the secondary carinal cartilages which were described for the first time and which were found to be the corner-stones of the terminal bifurcations of the main bronchi. Often fused to main and lobar cartilages they behaved like rings in front, but at the back they were always free and behaved like plates, showing a hybrid behaviour. It was felt that the importance which was attributed to them was not exaggerated as they proved to be the key to the never yet described cartilaginous structure of the bifurcations at the end of the main bronchi.

By ignoring these cartilages J. Hayward and L. Reid (1952) failed to see the differerence between the upper and lower branches of the left main bronchial bifurcation in the only contribution to the cartilaginous architecture of this part of the bronchial tree. In the preparation which served to illustrate their paper the last main bronchial rings had merged into one large plate, a situation which was met only once in our series (Fig. 23 E).

In the drawings by F. Merkel (1902) and A. Rauber and F. Kopsch (1955) secondary carinal cartilages were either absent or only shown on the anterior view. These pictures also included other inaccuracies so that no more than a historical value could be attributed to them.

Comparison of the five lobar bronchi, as far as their cartilaginous structure was concerned, revealed that they were all five different, but that a striking likeness existed between the lower lobe bronchi on the one hand and the right middle and left upper lobe bronchi on the other, while the right upper lobe bronchus had a cartilaginous support of its own.

1. The Right Upper Lobe Bronchus

In certain cases supported by rings, in the others by plates this bronchus did in some ways behave like right middle and left upper lobe bronchi. Unlike them it did have a membranous part and a secondary carinal cartilage was never found supporting its junction to the main bronchus. Like the tracheal or primary carina, the secondary carina belonging to the right upper lobe bronchus always carried a main bronchial ring and these two bifurcations could therefore be considered as possessing a similar cartilaginous support.

2. The Right Middle and Left Upper Lobe Bronchi

Following characteristics were found to be comparable in the cartilaginous skeleton supporting these two bronchi.

a) Both were supported in the great majority of cases by the same kind of rings as the main bronchi or modified rings which fused to the main bronchial ones in the same percentage of cases on both sides.

b) Both lacked a membranous part.

c) Both contained a typical secondary carinal cartilage which behaved in the same way: the posterior leg always being free and the anterior leg either free or fused to neighbouring cartilages in the same proportion of cases.

3. The Lower Lobe Bronchi

These bronchi were supported by cartilage in an almost identical way.

a) Both had an all-round support of plates and tertiary carinal cartilages which never fused to either main, right middle or left upper lobe cartilages.

b) Both were connected to the main bronchi by a fibrous attachment running obliquely.

4. Conclusion

As far as their cartilaginous framework was concerned, the main bronchi were finally seen to bifurcate at their termination in the same way into two branches of unequal importance, the lower lobe bronchi appearing as the off-shoots of a stem ending in the middle lobe bronchus on the right, in the upper lobe bronchus on the left. The way the right upper lobe bronchus was related to the main bronchus had more in common with the tracheal bifurcation.

As a result of these observations the theory could be put forward that the left upper and right middle lobe bronchi are but artificial entities and in fact part of the main bronchi. Acceptance of this way of reasoning would however entirely upset traditional views on bronchial anatomy and has therefore been rejected as premature.

IX. The Pulmonary Hilum

1. Rings and Plates

All along the preceding description cartilages were encountered which, in agreement with the accepted terminology were either rings, modified rings, plates and carinal cartilages. If modified rings and rings which had the same properties could be considered under the same heading, the same could be said of plates and tertiary carinal cartilages, provided the latter ones were assimilated to modified plates adapted to the bifurcation of tertiary or segmental bronchi. Rings and plates were ultimately the two different shapes in which the cartilages appeared in the bronchial tree.

2. Where Rings and Plates Were Found

Putting the parts of the bronchial tree which were analytically described in the preceding chapters together again and taking their cartilaginous skeleton as a whole into consideration rings and plates were found to belong to two distinct territories.

Rings were confined to the central part including the trachea, the main bronchi, the antero-superior wall of the right middle and left upper lobe bronchi and in some cases the proximal part of the right upper lobe bronchus.

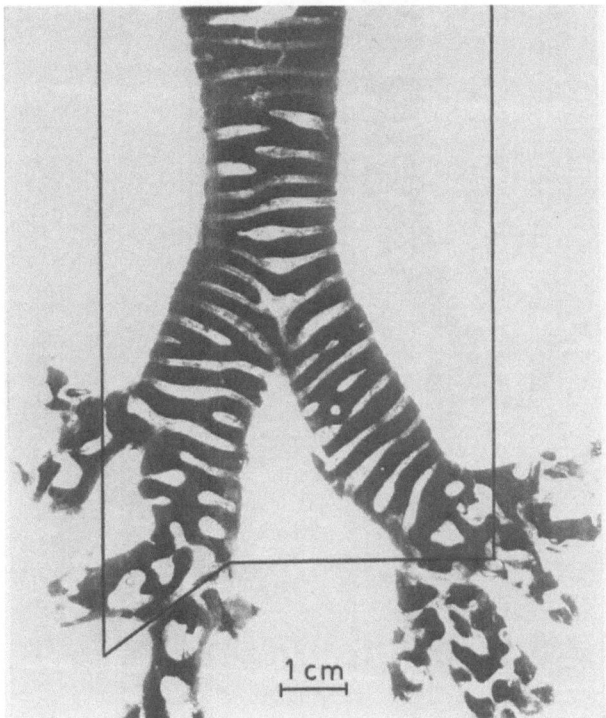

Fig. 28 A. The boundaries between the ringed and the plated parts of the bronchial tree.
Anterior half

Owing to the size, the shape and the capacity of rings to fuse, which grew more intense as the main bronchi grew longer, this central part was found to be a stiff, solid and rigid apparatus.

Plates belonged to the territories beyond: all the segmental bronchi *and* the two lower lobe bronchi. Airways in this region were pliable, faccid and elastic.

This difference in consistency was striking even in fresh preparations in which the upper segmental, but more surprisingly the lower lobe bronchi were found to be dangling from the central part as they hinged on their fibrous attachment.

3. Extra- and Intra-Pulmonary Airways

Outlining the territory supported by rings it did look at first sight that it corresponded to the extra-pulmonary territory. But why did it appear much larger in front than at the back (Fig. 28 A and B), and what about the lower lobe bronchi which were lying outside it and how precisely did the airways penetrate into the lungs ?

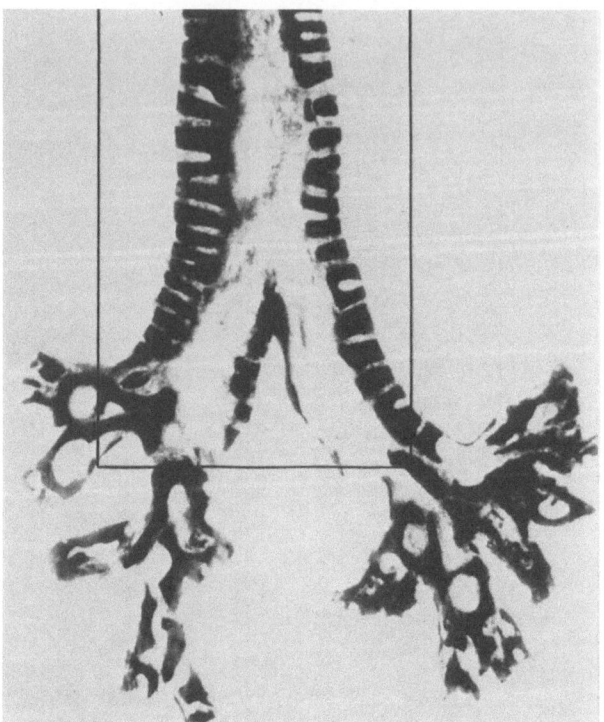

Fig. 28 B. Posterior half

The classical view vaguely is that the extra-pulmonary part of the bronchial tree reaches as far as the bifurcations of the main into lobar bronchi and to illustrate this pictures are produced in which the lungs have been severed from the hilar structures along a sagittal plane, showing the facies medialis of the lungs and the bronchi, the arteries and veins entering them in a perpendicular plane.

With a little more precision N. P. D. Smyth (1949) meant that the proximal part of the upper and middle lobe bronchi were outside the lungs.

These descriptions did not answer our questions with sufficient accuracy and in order to establish exactly which part of the bronchial tree was outside the lungs and which one inside, a series of 30 whole lungs were treated in the manner described page 13. Following were the results.

4. The Extra-Pulmonary Airways

In front airways seen to run outside the lungs were the trachea, the main bronchi, the proximal part of the right upper lobe bronchus and the greater part of the right middle and left upper lobe bronchi (Fig. 29 A).

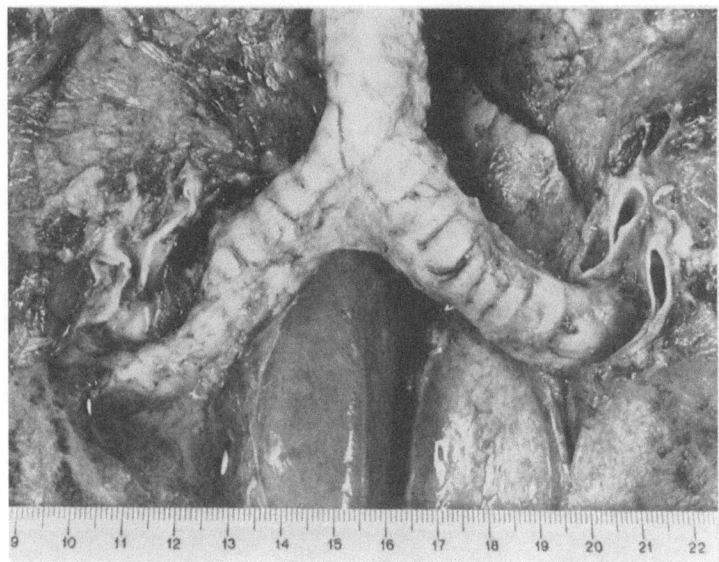

A

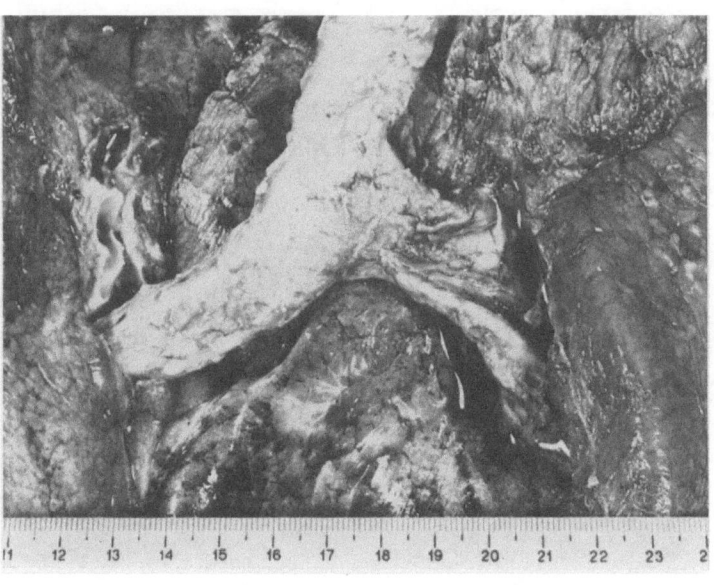

B

Fig. 29 A and B. Prep. H7, the extra-pulmonary airways. A Anterior view; B posterior view

4*

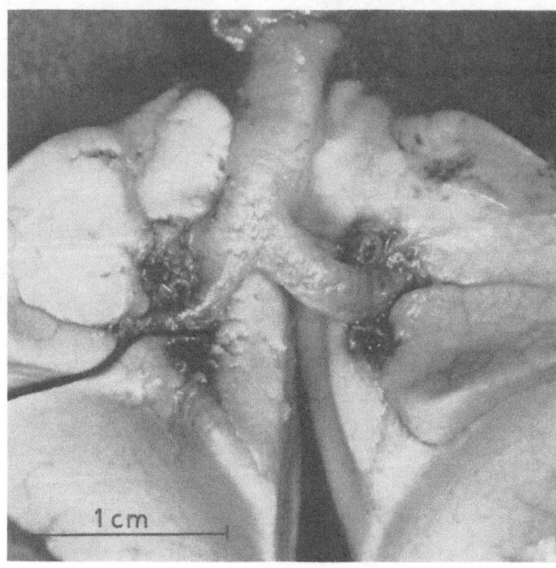

Fig. 30. Foetus No 76, C. L. R. 12.5 cm, aged 17 weeks, anterior view of extra-pulmonary airways

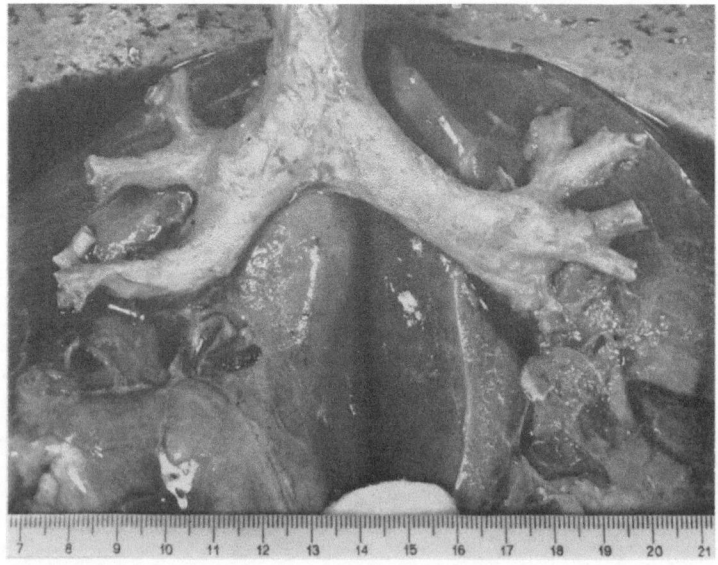

Fig. 31. Prep. H 19, the parenchyma of the upper and middle lobes being removed, the lower lobe bronchi are seen to enter the lungs in an oblique plane, they are immediately and entirely inside the lungs

In adults the boundary between extra- and intra-pulmonary territories was sometimes a little vague due to the large vessels surrounding them and on the left side part of the lower lobe bronchus became denuded if dissection was pushed too far. In foetal lungs however in which neither vessels nor lung substance had expanded, limits were sharp (Fig. 30).

Posteriorily the demarcation line where the airways entered the lungs was sharper, vessels being absent (Fig. 29 B). The extra-pulmonary territory was much smaller than in front meaning that the airways entered the lungs at a higher level, in other words in an oblique plane. No lobar bronchi were apparent, except the root of the right upper lobe bronchus. This lobar bifurcation was therefore the only one to lie entirely outside the lungs, like the tracheal bifurcation.

5. The Intra-Pulmonary Airways

Neither from the front nor from the back could lower lobe bronchi be seen on whole lungs. That they were lying entirely inside the lung substance was clearly demonstrated when the parenchyma from the upper and middle lobes was removed from the bronchi and the lower lobes only remained (Fig. 31).

Not only did this reproduction show that the course of the lower lobe bronchi was immediately and entirely intra-pulmonary, but also that the other lobar bronchi split into segmental branches as soon as they entered the lungs.

6. Cartilages Outside and Inside the Lungs

After preparations were stained and dissected it could be ascertained that rings ceased and plates began to appear precisely along the well demarcated line where airways entered the lungs. It was not possible to show this photographically, but by comparing pictures of halved stained specimens to the whole unstained preparations (Fig. 32 A and B and Fig. 29 A and B) rings were seen to correspond to the extra-pulmonary part and plates to the intra-pulmonary part of the respiratory apparatus.

7. Conclusions

This investigation on whole lungs, notwithstanding its simplicity clarified a number of findings which had appeared unusual and unexpected on stained bronchi.

a) It demonstrated how the transition from rings into plates coincided exactly with the entrance of the airways into the lungs.

b) It did show how the right upper lobe bronchus was the only one to originate outside the lungs and penetrate into it in a sagittal plane, explaining why it had a membranous part, lacked a secondary carinal cartilage and had to be assimilated to the tracheal bifurcation.

c) It made it clear how the right middle and left upper lobe bronchi penetrated obliquely in their lobes, leaving the anterior and superior, ring-supported sides outside, and their posterior, plate-supported wall, inside the lung substance.

d) It also showed how the secondary carinal cartilages possessed hybrid characteristics, behaving in front like rings and at the back like plates (p. 48).

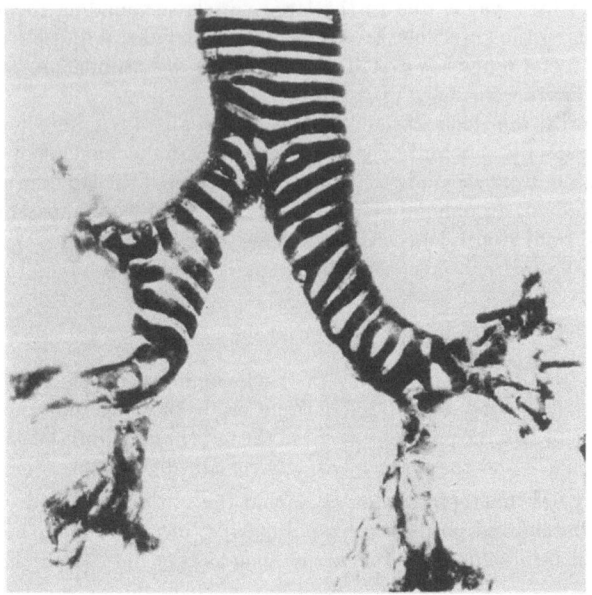

Fig. 32 A. The stained cartilages of prep. H 7, to be compared to Fig. 29. Anterior half

e) It explained how the intrapulmonary lower lobe bronchi emerging obliquely from the main bronchi were supported from the start by cartilages of intra-pulmonary morphology.

f) It finally allowed us to visualize better why the apical lower segmental bronchus, an off-shoot of the lower lobe bronchus usually emerged opposite, or even higher than the right middle or left upper lobe bronchus.

X. Foetal Studies

Pursuing our studies a series of foetal lungs were submitted to the macroscopic staining method in order to investigate how it could contribute to our knowledge of the histogenesis of tracheo-bronchial cartilage.

1. Embryology

Embryologically the airways arise on the anterior wall of the fore-gut as the laryngo-tracheal bulge, where chemical phenomena such as accumulation of glycogen, appearance of enzymic activities, noticeably that of alkaline phosphatase and increase of cytoplasmic ribonucleoprotein in the epithelium precede macroscopic morphological changes (S. Sorokin, 1965).

Once differentiated, the bulge separates from the oesophagus by ingrowing septa, leaving the cranial primitive laryngeal aditus as the only communication

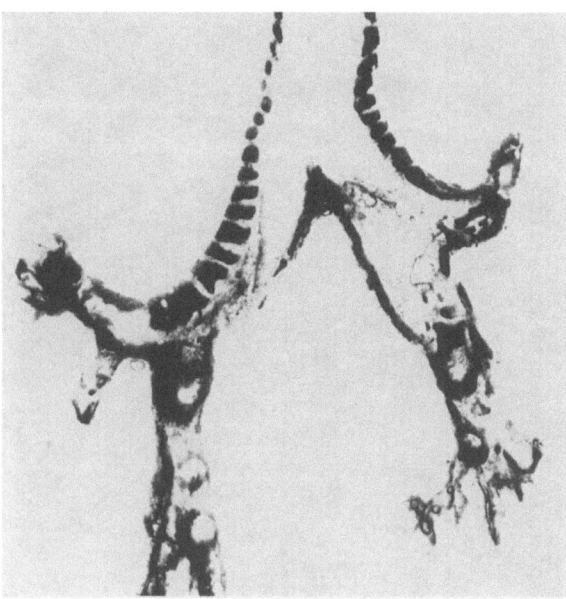

Fig. 32 B. Posterior half, for comparison with Fig. 29 B the picture has been reversed from left to right

between oesophagus and trachea. But the end of the trachea already splits into two primitive lung buds (one should say main bronchial buds) which grow not only caudally and laterally, but also dorsally, embracing the oesophagus like a fork (R. His, 1887).

The five lobar buds, three on the right and two on the left arise from the main bronchial buds. By a burst of activity (E. A. Boyden, 1955) in segmental branches, they develop into whole lungs in the pleural space.

V. E. Krahl (1964), using information already acquired (B. M. Patten, 1948) drew our attention to the fact that the earliest stage of the development of the airways from the fore-gut is situated below the heart, but finally comes to lie above it. The development of the trachea and the main bronchi therefore takes place in the mesenchymal mass between the heart and the vertebral column which is due to become the mediastinum. These facts had also been illustrated by G. L. Streeter (1945, 1948).

The arborization of lobar into segmental and smaller branches on the other hand develops into the pleural space, which is separated from the coelomic cavity by the ingrowing pleuro-pericardial folds.

2. Histogenesis of Cartilage

Cartilage appears much later. During the 7th week after L. B. Arey (1954) and the 8th week according to A. Fischel (1929) and W. J. Hamilton (1952).

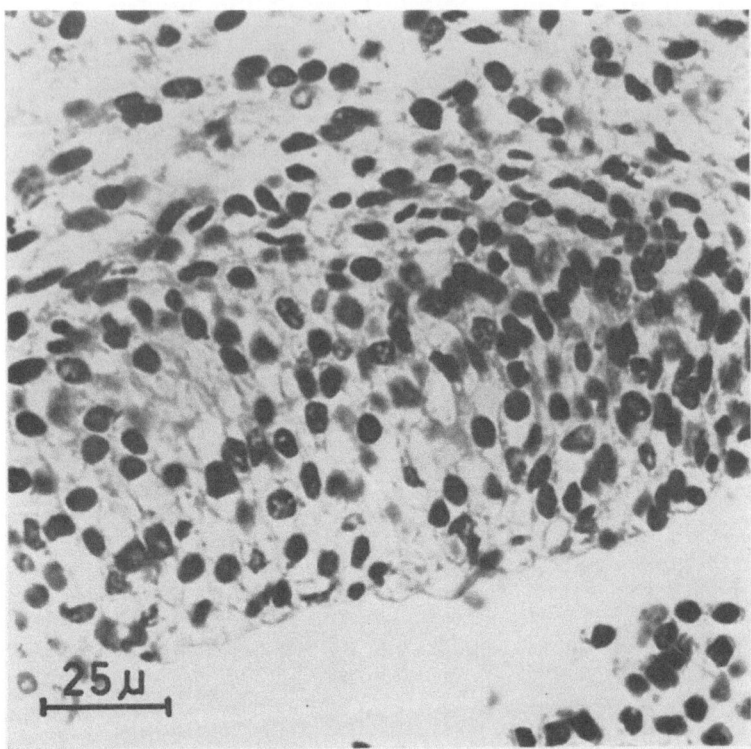

Fig. 33 A. The differentiation of cartilage in the airways. Prep. F3, 60 days, section of the
extra-pulmonary airways in the precartilage stage

U. Bucher and L. Reid (1961), using histological sections and various staining
methods, described the earliest appearance of precartilage in the main bronchi
during the 10th week and ground substance between the 12th and 14th week.
They noted the appearance of plates in the intra-pulmonary airways in steady
progression from the hilum to the periphery without any increased activity at any
special age.

I. Brenek (1941) in a reconstructed bronchial tree of a foetus 50 mm CRL,
aged ± 11 weeks, found differentiated rings in trachea and main bronchi. In the
right one she counted eight and found connections at the tracheal bifurcation and
the right upper lobe bronchus. Nuclei of carinal cartilages were differentiated in
bifurcations of second and third order. Measuring the distance between the trachea
and the differentiated plates farthest away from it, she concluded that the left
lung grew faster than the right one.

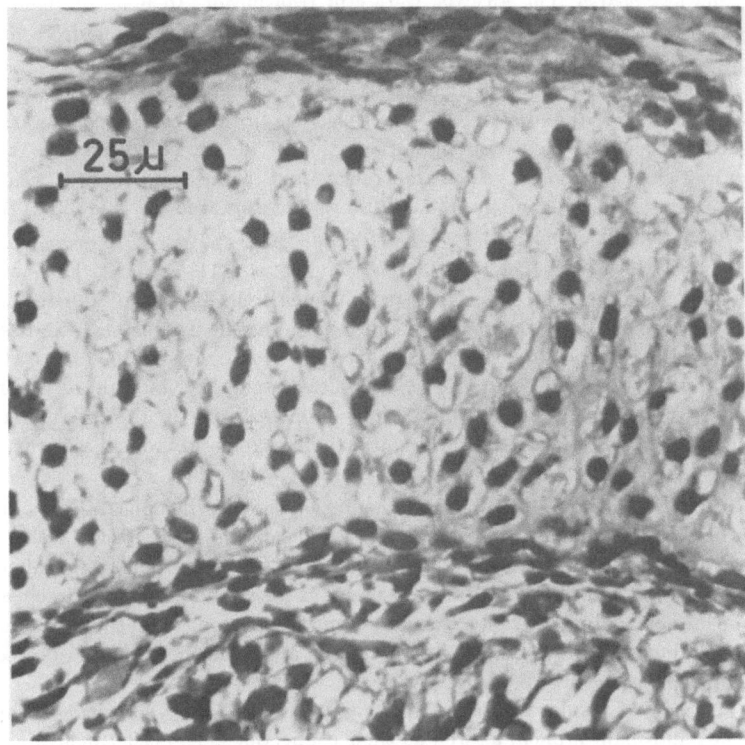

Fig. 33 B. Prep. F 11, 80 days, the appearance of ground substance in the extra-pulmonary airways

3. Own Investigation

Application of the macroscopic staining method to investigate this aspct of tracheo-bronchial cartilage was found to be much more delicate than in adult preparations. Maceration meant failure and at the critical period when cartilage appears, absence of staining could as well be due to chemical changes in the cells as to absence of cartilage. Which explains why out of 87, only 30 preparations were accepted as being satisfactorilly stained (Table 1). Observations were divided into three distinct periods.

a) First Period. 7th to 11th Week. Foetus F 1, 50 days. The preparation was coloured pale green but the terminal bronchioli somewhat darker, giving the lungs a grape-like appearance. The airways outside as well as inside them were completely translucent and could not be distinguished inside the lungs. No stained cartilage was seen.

Foetusses F 2, 58 days, F 3, 60 days, F 4, 63 days, F 5, 68 days, F 6, 69 days, F 7, 73 days, F 8, 74 days, F 9, 77 days and F 10, 77 days. All these preparations were more or less pale green, but as they grew older, their grape-like appearance became less coarse. The airways were still translucent. In none of them could any stained cartilage be distinguished. No separation of lung tissue from bronchi by dissection was possible in any of these lungs.

Preparations F3, 60 days, F6, 69 days and F10, 77 days were submitted to histological sections and stained for cartilage. In all three the shape and arrangement of the chondrocytes and the orientation of the perichondrial cells, typical of precartilage were observed in the trachea and in the main bronchi. Chondrocytes were swollen, but no ground substance had appeared yet (Fig. 33 A). Inside the lungs no appearance of precartilage was noticed.

b) Second Period. 12th and 13th Week. Foetus F11, 80 days. This was the youngest preparation in which lung parenchyma could be carefully dissected free from the bronchi, at least in the vicinity of the hilum. It was also the first one in which the rings in the extra-pulmonary part were lightly stained. Also stained were the nuclei of the secondary carinal cartilages, one nucleus of a tertiary carinal cartilage in the right upper lobe bronchus, one in the left upper lobe bronchus, between the apico-dorsal and lingular segmental bronchi and one in the orifice of the apical lower segmental bronchus on the right.

Histological examination of this preparation showed the appearance of ground substance in the macroscopically stained cartilages and nuclei (Fig. 33 B).

Foetus F12, 82 days and foetus F13, 84 days. Staining of all extra-pulmonary cartilages was still discrete. As in the previous foetus staining of all tracheal and main bronchial rings was of the same intensity, meaning that they was no progression, but that they stained simultaneously and that there was no precedence of one main bronchus over the other.

Foetus F14, 87 days. Cartilages were coloured brightly showing all rings in trachea and main bronchi as well as their connections. The last tracheal and the first main bronchial were seen to be forked, ten rings were counted in the right and ten in the left main bronchus. The two secondary carinal cartilages were completely stained as well as their connections to the main bronchial rings.

Cartilages stained in the lobar branches were:

Right upper lobe bronchus: one ring fused to the upper part of the main bronchus and two tertiary carinal cartilages.

Left upper lobe bronchus: one free ring and three tertiary carinal cartilages.

Right middle lobe bronchus: only the tertiary carinal cartilage.

The lower lobe bronchi: the lower cartilages in the orifice to the apical lower segmental bronchus, three tertiary carinal cartilages in the bifurcations into basal segmental bronchi and on the right two cartilages in the orifice to the cardiac segmental bronchus (Fig. 34).

Foetus F15, 88 days. All the cartilages mentioned in the previous preparation were stained. Nuclei of quaternary carinal cartilages appeared in all lobes: four in the right upper, four in the left upper, one in the right middle, five in the right lower and six in the left lower lobe bronchus. In the latter nuclei of plates were already making their appearance (Fig. 35).

In foetusses F16, 89 days and F17, 90 days, findings were similar.

Foetus F18, 90 days. In this anomalous lung an azygos lobe contained one ring, a tertiary carinal cartilage and four small plates. The right upper lobe

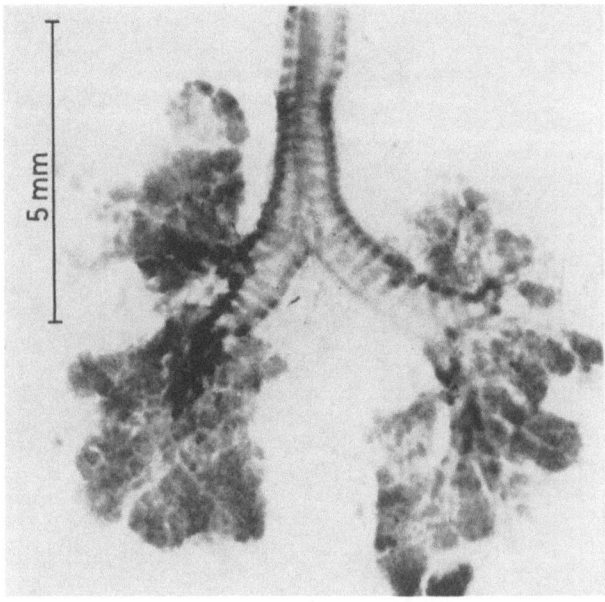

Fig. 34. Prep. F 14, 87 days, for stained intra-pulmonary cartilages see text

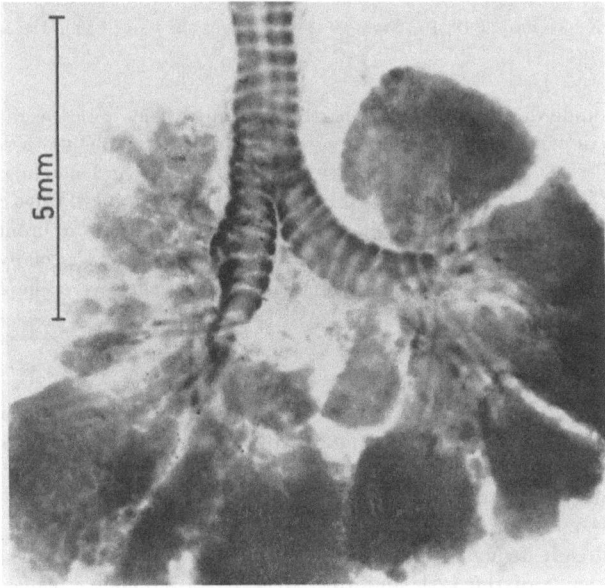

Fig. 35. Prep. F 15, 88 days, description in text

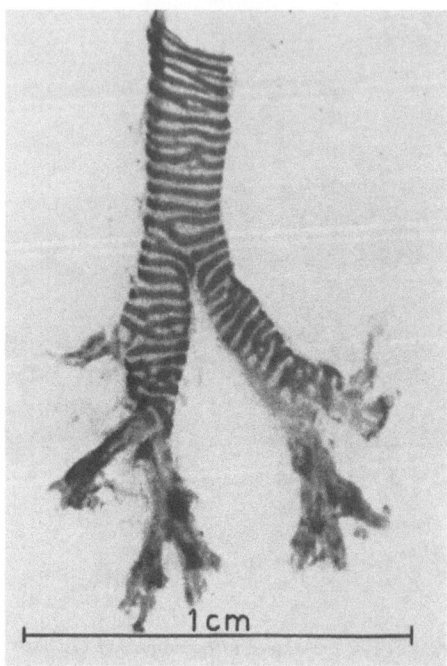

A

Fig. 36 A and B. Prep. F 19, 102 days, all the cartilages are stained as in the adult

bronchus was underdeveloped and no cartilage stained in it. For the remainder differentiation of cartilage was as far advanced as in the previous foetusses.

c) Third Period: 14th Week and Older. In all preparations belonging to this period (F 19—F 30) all cartilages involved in this study were stained: rings could be counted, connections recognized. Dissection of parenchyma from bronchi was easy but halving for reproduction met with some difficulty, especially where smaller bronchi were concerned. Characteristics of cartilages were the same as in adult bronchial trees (Fig. 36).

4. Summary and Conclusion

Macroscopic staining of foetal lungs proved its usefulness by confirming previous observations by I. Brenek (1941) and U. Bucher and L. Reid (1961) obtained by other methods and provided a clear picture of the histogenesis of tracheobronchial cartilage in man.

a) Before the 11th week no cartilage stained although the presence of precartilage was already demonstrated histologically in the extra-pulmonary airways.

b) In the course of the 12th and 13th week extra-pulmonary rings stained simultaneously in trachea and main bronchi, no difference being noticed between

B

the two main bronchi. At the same time the nuclei of the secondary carinal cartilages appeared.

c) Nuclei of tertiary and quaternary carinal cartilages stained progressively and radially from the hilum to the periphery.

d) Progression of differentiation in the segmental bronchi was not exactly harmonious. It did occur somewhat faster or slower in one or other segmental bronchus depending on the individual, agreeing with the somewhat erratic growth of segmental bronchi (R. Heiss, 1919).

e) The fact that the lower lobe bronchi and their branches showed a greater amount of cartilage earlier than the other lobar bronchi and their branches was due to the fact that they possess a greater number of tertiary carinal cartilages.

f) Progression of differentiation of cartilages did not show the same burst of activity as does bronchial branching between the 10th and 14th week (E. A. Boyden, 1955; U. Bucher and L. Reid, 1961).

g) From the 14th week onward all cartilages involved in this study were completely stained as in the adult.

h) Generally speaking the histogenesis of the cartilages belonging to the extra-pulmonary airways which had developed in the mesenchymal mass of the media-stinum was simultaneous, opposed to the histogenesis in the peripheral airways which had developed in the pleural spaces and which was progressive.

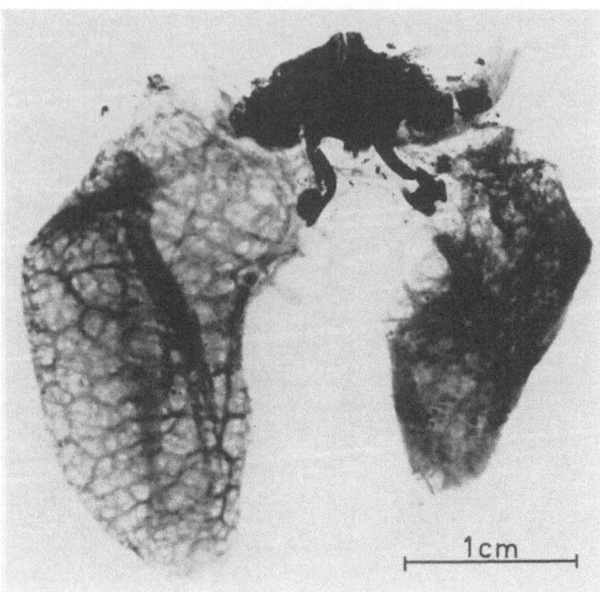

Fig. 37 A. In the toad the laryngeal and crico-tracheal cartilages are brightly stained, there is neither trachea nor intra-pulmonary cartilage

XI. Age and Sex

Of 47 preparations age and sex were known as shown in the illustrations.

A. Sex. Twenty one male bronchial trees were compared to twenty one female ones. All characteristics of cartilages as described in this study: type, shape, size, situation, relationship to other cartilages and their percentages were the same in both sexes.

B. Age. A group of 21 preparations belonging to individuals ranging from 35 to 61 years, average 51 years was compared to a group of 21, aged between 67 and 89 years, average 75 years. No change in type, shape, size of cartilages, nor the way they were related to others were observed, confirming G. Benek *et al.* (1966) observations on tracheal cartilages.

The part of the cartilaginous skeleton of the tracheo-bronchial tree which is completely developed at the intra-uterine age of 14 weeks does therefore undergo no change until death.

XII. Animal Studies

Much information about tracheo-bronchial cartilages in animals has been collected by J. F. Meckel (1833), G. Cuvier (1840) and other comparative anatomists. Chr. Aeby (1880) and A. Narath (1901) who belonged to them also in-

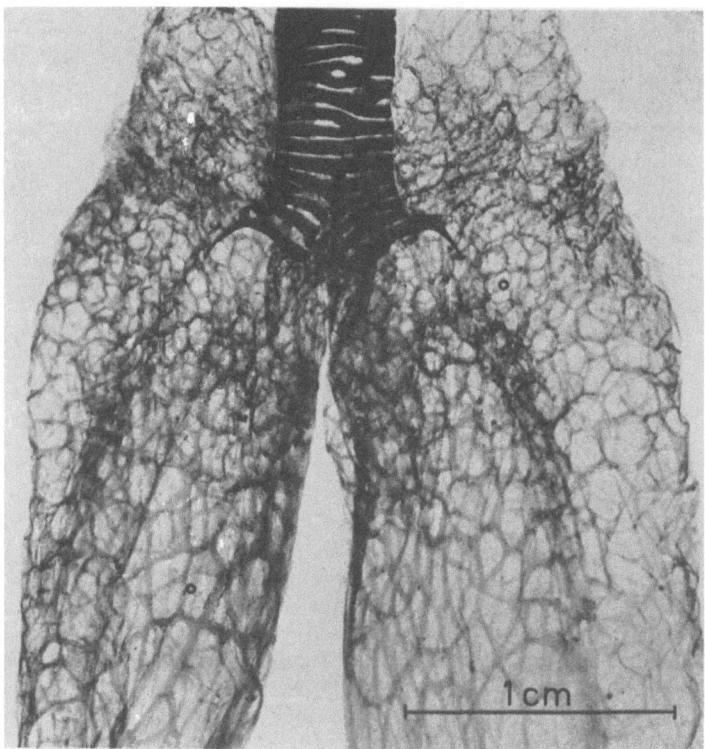

Fig. 37 B. Extra-pulmonary rings and symmetrical main bronchi in chameleon

cluded a number of animal species in their studies on the bronchial tree, but they completely disregarded the existence of cartilage. The reason, as we have said before, being that they used casts or corrosion preparations of the airways in which the cartilage containing walls had disappeared. More recent contributions were made by J. E. W. Ihle *et al.* (1924), H. Marcus (1937), C. K. Weichert (1958) and others.

From the amphibia onward development of the neck goes together with the interposition of the trachea and the main bronchi between the larynx and the lungs. Absent in the frog it reaches large proportions in snakes, giraffes and some birds and the number cf cartilaginous rings varies accordingly.

Inside the lungs cartilage may be absent (Anura, Aphidiae, Aves) or not (Gymnophiona, Pipa, Urodela, Crocodilia and Mammalia) H. Marcus (1937). Amongst mammals it is lacking in Mycetes and little developed in Marsupialia, Rodentes and Simiae (J. E. W. Ihle *et al.*, 1924). In men, intra-pulmonary cartilage has been found to be less developed in Mongols than in Europeans (W. Stefko, 1931).

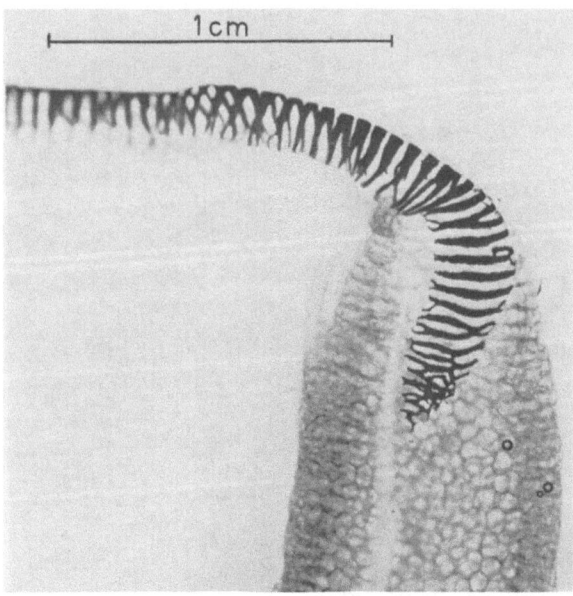

Fig. 37C. In the snake the lung which does not show cartilage is seen as a bulge of the membranous part

About the transition between extra- and intra-pulmonary cartilage, information was less precise and since the macroscopic staining method showed it much clearer than could be expected from any other method, a number of animal preparations were stained and showed following results.

1. Amphibia

In the *frog* and the *toad* the trachea and the main bronchi were absent. The laryngeal and crico-tracheal cartilages were conspicuous and inside the lungs septa could be seen but no stained cartilage (Fig. 37 A).

2. Reptilia

The trachea of the *chameleon* was supported by single and compound rings and a membranous part. From the observation of two preparations it could not be decided whether each main bronchus contained four or five rings. As in man it depended which ring was considered the last tracheal one and no clear boundary existed between the two parts. Inside the lungs the airways were but primitive air sacks, separated by septa, the walls of which did not contain cartilage (Fig. 37 B).

In the *snake* (Coronella), the very long trachea contained single and compound, mostly forked cartilages which surrounded only half of its circumference,

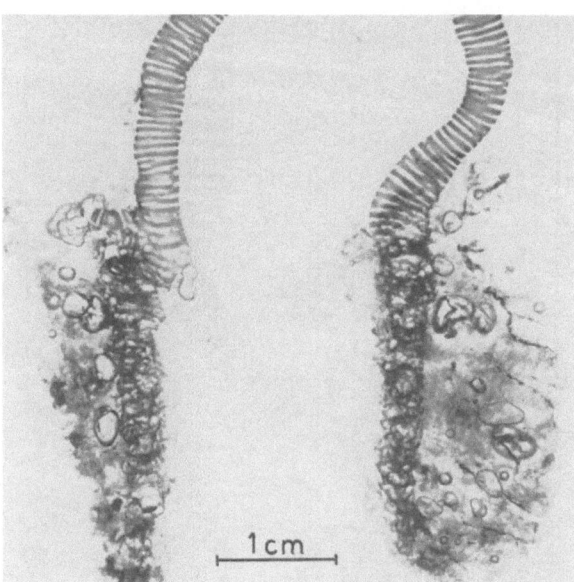

Fig. 37 D. In the tortoise an intra-pulmonary bronchus is supported by cartilages which are differently shaped than the main bronchial ones (round shadows are air-bubbles)

leaving a broad membranous part. There was no main bronchus, and where the trachea met the only lung, rings gradually shortened until they disappeared completely. The trachea was seen to lie on the lung for some distance and the latter appeared as a large bulge of the membranous part. No cartilage was seen inside the lung (Fig. 37 C).

The rings in the short trachea of the *tortoise* were interrupted at the back but in the long main bronchi they were complete. Forked and single rings were found. Where the airways entered the lungs cartilages ceased to be regular rings and became an irregular network of cartilage surrounding one important bronchus crossing the lung from above to below. Shortly after this airway had entered the lung it showed a bulge as if another bronchus leading to the apex originated there. Small bits of cartilage were seen to be scattered here and there about the lung parenchyma (Fig. 37 D).

3. Aves

In birds neither the syrinx nor the intricate system of air sacks were taken into consideration.

The trachea and the two main bronchi of the *lark* were surrounded by complete, single rings: a membranous part was lacking. On entering the lung substance rings became thinner, shorter and more widely spaced and on the original preparations it could be seen that cartilages ceased exactly where the main

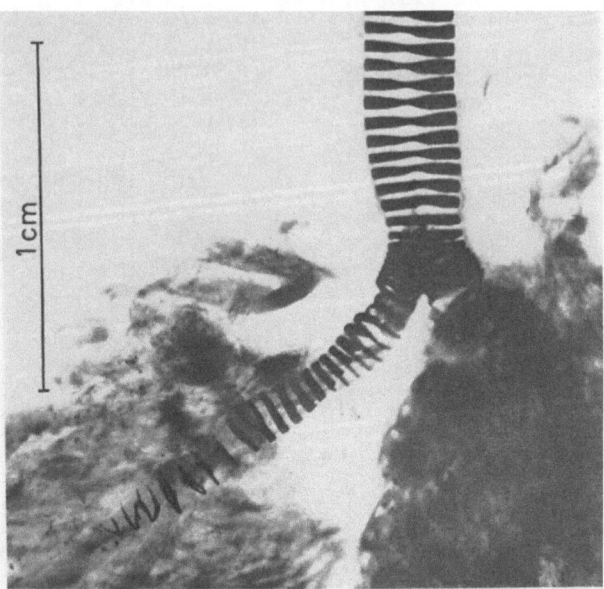

Fig. 38. In birds (lark) the main bronchial rings cease where the meso-bronchus begins, to show
this the parenchyma masking this transition has been dissected away on one side

bronchi became mesobronchi. To show this on a reproduction, most of the dense
parenchyma had to be removed (Fig. 38).

4. Mammalia

In the *dog* the right lung had four, the left one three lobes but the cartilaginous skeleton of its airways could be compared with that of man. Trachea and
extra-pulmonary main bronchi were supported by broad rings which almost
touched each-other at the back, leaving but a small paries membranacea. In each
main bronchus five rings could be distinguished. Lobar and segmental bronchi,
on the contrary, were completely surrounded by a fine network of cartilage. The
transition between the two types of cartilage was sharp and coincided with the
entrance of the airways into the lungs (Fig. 39 A and B).

The lungs of the *rat* had but one lobe on the left and four on the right side.
In stained preparations ten half rings were found in each main bronchus. They
were single, had the same appearance as in the trachea and were completed at the
back by a broad membranous part. In the right upper lobe bronchus one ring
was found. The main bronchial rings ceased as the airways became lung on the
left and middle lobe on the right, and as the lungs themselves did not contain
cartilage, the demarcation between extra- and intra-pulmonary airways was sharp
(Fig. 40). In the mouse findings were similar.

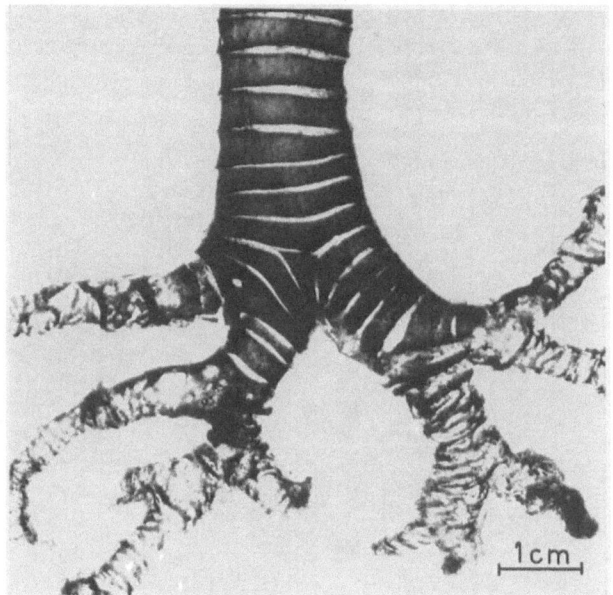

A

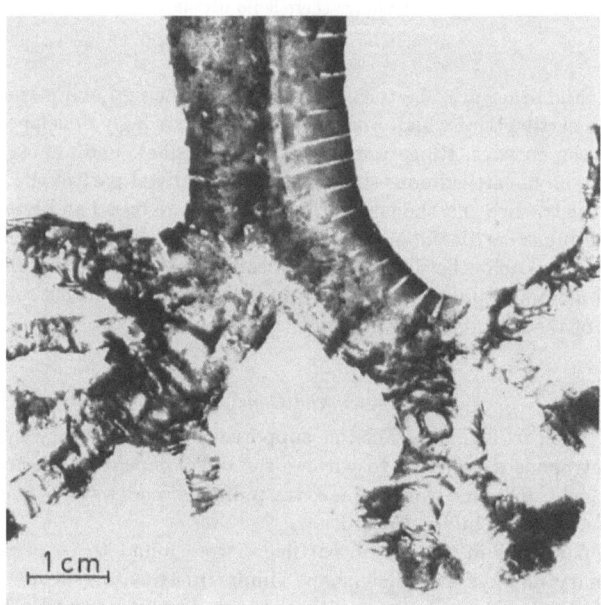

B

Fig. 39 A and B. The cartilaginous skeleton of the bronchial tree of the dog. A Anterior half;
B posterior half

5*

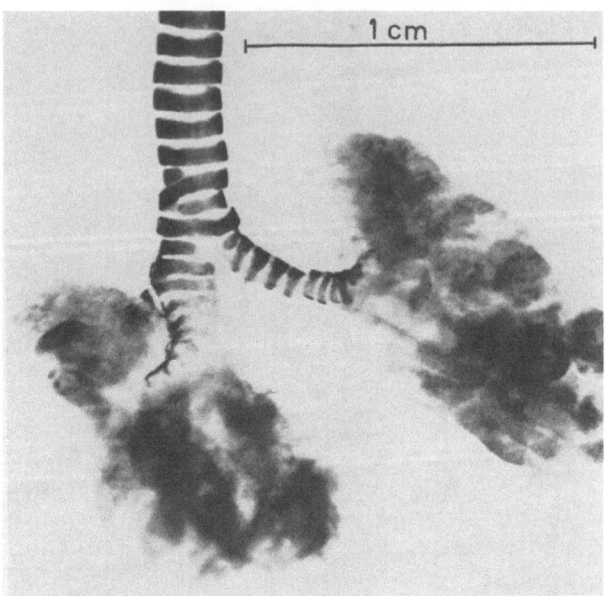

Fig. 40. The rat has an extra-pulmonary cartilaginous skeleton which resembles that of man.
Inside the lungs there is no cartilage

In the *calf* and other artiodactylae (the *sheep* and the *pig*) and perissodactylae
(the *horse*) the cartilages of which were investigated a strongly developed skeleton
was found in the airways. Rings were bulky, mostly single and almost complete.
In the horse, a large cartilaginous shield covered the distal part of the trachea at
the back. In the trachea and the main bronchi rings were broad and lying parallel,
but inside the lungs cartilaginous structures were more slender, unevenly distributed and joined to each-other like a network completely surrounding the airways.
Also in these mammals did the change in morphology of cartilages coincide with
the entrance of the airways into the lungs (Fig. 41 and 42).

5. Summary and Conclusions

Study of the cartilaginous skeleton supporting the respiratory system in a
number of tetrapods enabled us to witness the development of the trachea and
the main bronchi, in other words, the extra-pulmonary airways between larynx
and lungs from the amphibies onward.

In all preparations in this series, cartilages were found to be present in the
extra-pulmonary airways and their shape similar in trachea and main bronchi,
whereas inside the lungs cartilage was either absent (amphibia, reptiles, birds, rats)
or shaped otherwise and distributed differently (other mammals). In the tortoise
a transitional situation was found.

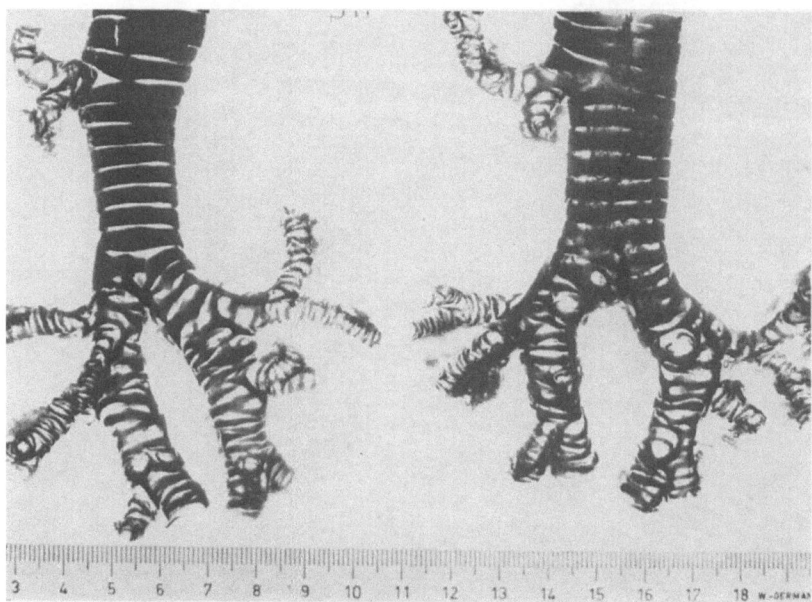

Fig. 41. Anterior and posterior halves of a stained sheep's bronchial tree

The boundary between the cartilaginous skeleton outside and inside the lungs was always sharp and coincided with the entrance of the airways in the lungs.

The morphology of cartilages in trachea and main bronchi being similar, the transition between the two was ill defined.

All these findings were in agreement with our observations in man and with the development of cartilage in foetusses in which all extra-pulmonary cartilage stained simultaneously, confirming a statement by H. Marcus (1937), that "extra-pulmonaler Bronchus und Trachea sind also gleiche Gebilde, und wo eine Lunge, wie bei Schlangen völlig rückgebildet ist, sagen wir, daß die Luftröhre direkt in die Lungen übergeht, also ein extrapulmonaler Bronchus fehlt".

Consequently the old way of calling the main bronchi: "ramus dexter et sinister tracheae" which implied the structural identity of these parts was more appropriate than the modern appellation: "bronchus principalis dexter et sinister", based on convention, tradition and confirmation by nomenclature commissions. Structurally and morphologically it would be preferable to reserve the term "bronchus" to the intra-pulmonary airways.

As the cartilaginous rings were seen to continue in the main bronchi in mammals as well as in other species, and cease at the entrance of the airways in the lungs, all evidence of a "Stammbronchus" (Chr. Aeby, 1880) reaching from the tracheal bifurcation to the extreme periphery of the lung base was seen to fall to pieces as our study progressed.

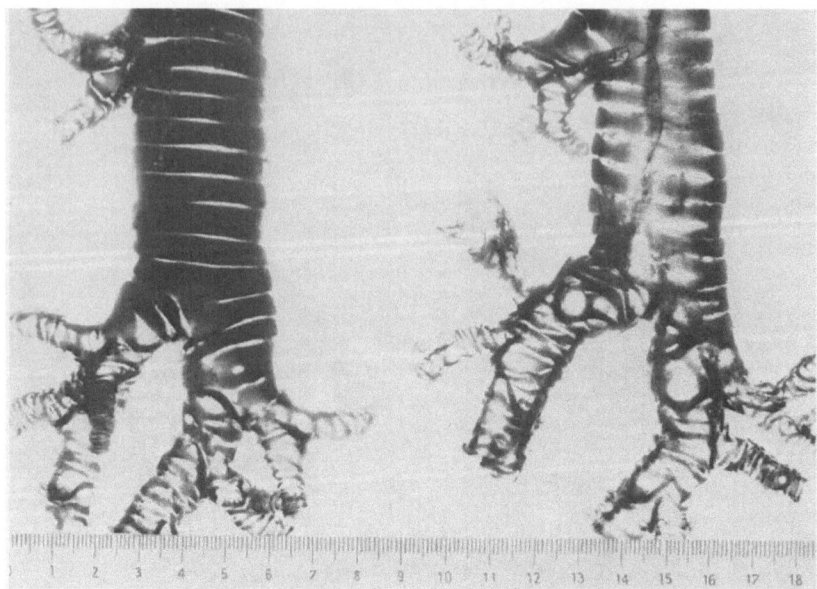

Fig. 42. Anterior and posterior halves of a calf's cartilaginous skeleton

The symmetry between the two main bronchi in reptiles and birds, less conspicuous in untreated human preparations, became again apparent when their cartilages were stained and their rings counted.

The mammalian and also the human respiratory apparatus could therefore be visualized, when stained, as the two-lunged system of the reptiles to which a third lung; the tracheal or the right upper lobe had been added.

In which case, the lateral bulge of the intra-pulmonary airway at the apex of the turtle's lung could be considered as the future middle lobe bronchus on the right and the upper lobe bronchus on the left.

Our findings did indeed undoubtedly show that in mammals and this was especially clar in rat, dog and man, the cartilaginous main bronchi both divided at their end in the same manner: into a middle and lower lobe bronchus on the right and into an upper and lower lobe one on the left, confirming belatedly Chr. Aeby's (1880) theory of homology between right middle and left upper lobes.

The two branches being differently equipped with cartilage, this mode of branching could be considered one of uneven dichotomy as found by K. Horsfield and G. Cummings (1968) in peripheral airways in human lungs.

As to lobation, characteristic of mammals, there did not seem to exist any correlation between lobation and bronchial distribution: in rat and in man the left main bronchus divides into an upper and a lower branch but in the former there is one lobe, in the latter two.

Many more animal investigations will of course be required to assess the value of the above statements, but from now on it may be said, that never before has the human bronchial tree been shown to fit into a logical evolutionary system so convincingly as by taking into consideration its cartilaginous skeleton.

XIII. Morphogenesis

The problem of the morphogenetic dynamism which influences the different shape and situation of the extra- and intra-pulmonary cartilages arises now. But only hypothetical considerations can be ventured in the present state of our knowledge.

Comparative anatomists often maintain that the cartilages in the airways are homologous to the branchial arches in the fish (H. Braus, 1929; P. Brien and A. Dalcq, 1954). The two first branchial arches: the mandibular and the hyoid and part of the third would give rise to the viscero-cranium and the next greatly regressed ones would take part in the development of the airways. This view is partly based on the shape of the rings which in most species are open at the back and suggest the skeleton of two paired branchial arches joined in front by a copula. From the Amphibia onwards the lungs migrate from the larynx to the pleural space, due to the development of the extra-pulmonary airways, while the larynx and more particularly the cartilagines laterales of Necturus and the cartilagines crico-tracheales of Rana do seem to descend from the seventh branchial arch (J. E. W. Ihle et al., 1924).

The cartilaginous skeleton of the extra-pulmonary airways, whichever its phylogenetic meaning, does in any case show a definite segmental structure. The segments are too many to correspond to the original body segments within the region in which they lie. The question how these segments appear remains unanswered as does the problem of segmentation in general. The hypothesis of a physiological competition between metabolites has been proposed (S. Spiegelman, 1945): a slightly privileged region would attract metabolites from above and below. Consequently negative zones, the future intersegmental spaces, would appear and draw away competition from more distant areas, allowing new dominating zones or future segments to build up (A. Dalcq, 1949). This explanation is, of course, entirely conjectural.

More than once has been demonstrated that the chorda dorsalis, and to a lesser degree the neural plate, favour the histogenesis as well as the segmentation of the somites in Amphibia (S Hörstadius, 1944; M N. Ragozina, 1946; I. Kitchin, 1949; W. Muchmore, 1951; H. Chuang and M. Tseng, 1956; H. Takaya, 1956). This problem was also investigated in birds, but results were less convincing (B. Menkes et al., 1961). Only recently did R. Lanot (1967), working on the same material, conclude: "These factors plead for the existance of a diffusible factor which would on the one hand determine the appearance of successive somites and on the other allow the further differentiation of the somatic elements". Which corresponds to some sort of physiological competition. Whether the chord or the nervous system intervene in the appearance of metabolites (diffusible factor) may be disregarded as far as our problem is concerned.

A difference in morphogenetic potential (A. Dalcq, 1941) is bound to exist in any case in the walls of the airways, between the cartilaginous segments and the parts in between which do not contain cartilage. The many "faults" in the segmentation would be explained by the fact that this difference is slight.

The characteristic cartilaginous architecture of extra- and intra-pulmonary airways can hardly be explained by the morphogenetic influence of a different function, because the whole cartilaginous skeleton is completely formed in its final shape, in the young foetus, long before birth and the first act of breathing. More convincing is the difference in relationship between the mesenchyma which surround the entodermal airways in which the cartilages will develop on the one hand, and the pleura on the other in the two corresponding regions. Extra-pulmonary airways are indeed covered only front and sideways by pleura which continues along the sides of the oesophagus. But where the lung buds grow into the pleural spaces, this mesenchyma is covered on all sides by the pleura it pushes forward: the future visceropleura.

Mesenchyme, we know, is capable of exerting an inductory influence upon the parenchyma of different organs: submandibular gland, kidney, liver, skin. More in particular may pleural mesenchyme induce branching in the entoblastic epithelial airways (J. Alescio, 1962) and the question may be asked whether the close relationship of the pleural serous membrane may not influence the differentiation of this mesenchyme itself. In which case the explanation would be at hand why, in most species, the extra-pulmonary cartilages are interrupted at the back, the intra-pulmonary ones are circumferential and why the membranous part ends where the airways penetrate into the lungs.

Within the lungs more or less regular rings are no longer found, but either an irregular network as in dogs, calves and horses, or circumferential plates as in men, or no cartilages at all as in rats. What ultimately occurs could be the results of a conflict between a progressive factor (influence of pleura) and an inhibiting one. J. Fautrez (1964) has drawn our attention to the influence of inhibition upon harmonious development. He calls harmonious development as it occurs in nature, lacking any objective criterium. Whichever system is considered, this development remains between two boundaries. It is pushed beyond a lower level by extrinsic and intrinsic morphogenetic factors (A. Dalcq, 1941) but kept under an upper level by inhibiting forces. This could be applied to our system and one could imagine that lung tissue has an inhibiting influence, because regular cartilage appears outside the lung (where the mesenchyma lies next to the pleura) but not within it.

The final effect between the two factors would lie at a different level according to the species. And one can also imagine that contact between lung tissue and airway is less close, or less inhibiting, in the sharp corner between bifurcating bronchi, explaining why bifurcation cartilages develop earlier and become stronger than ordinary plates.

Summary

Methods used so far in the investigation of the bronchial tree having failed to produce a systematic study of its cartilaginous skeleton, a macroscopic staining method was employed in a series of 100 adult human, 30 human foetal and 12 animal specimens.

A detailed description of all cartilages supporting the human bronchial tree could thus be undertaken and for the first time the arrangement of cartilages around the lobar bifurcations and in the lobar bronchi explored.

Both main bronchi were found to contain roughly the same number of rings showing a symmetrical arrangement. Rings were also present as such in the left upper lobe bronchus, in a modified shape in the right middle lobe bronchus and in a high proportion of cases in the right upper lobe bronchus, whereas the lower lobe bronchi never contained rings, but circumferentially distributed plates.

The terminal bifurcations of the main bronchi into upper and lower lobe bronchi on the left and middle and lower lobe bronchi on the right were supported by an arrangement of cartilages which was in many ways the same on both sides, while the right upper lobe cartilages were more like the ones in the tracheal bifurcation.

The transition between rings and plates could be clearly established and was found to coincide with the entrance of the airways in the lungs, rings being found outside (extra-pulmonary) and plates inside (intra-pulmonary). The extent of these two territories and the way bronchi enter the lungs were clearly demonstrated.

In human foetusses the appearance of cartilage during the 12th and 13th weeks was confirmed, extra-pulmonary cartilage staining simultaneously and intra-pulmonary cartilage radially from the hilum to the periphery.

In animals the appearance of the trachea and main bronchi between the larynx and the lungs was witnessed and like in man a difference in morphology between extra- and intra-pulmonary cartilages consistently found, allowing a number of phylogenetic considerations.

Hypotheses on the morphogenetic dynamism which may influence the shape of cartilages were put forward.

Acknowledgement

The subject matter of this monograph, presented as a thesis at Ghent University (F. Vanpeperstraete, 1968) was originally written and printed in Dutch. But the present edition is not a mere translation: most chapters have been rearranged; cumbersome tables, which can be consulted in the thesis, omitted; superfluous illustrations dropped, but others, which may better show the many variations encountered in bronchial cartilages, added. It is hoped, that by doing so, we have been successful in presenting the subject in a more attractive way, likely to appeal to a wider range of readers.

We are firstly and mostly indebted to Professor J. Fautrez, head of the Department of Anatomy, University of Ghent, for making us share his enthousiasm for anatomy, for his advise based on experience and authority, his friendly criticism and his permanent encouragement in the course of our investigations, but above all for his sponsorship of our thesis and his help in preparing our manuscripts.

To Professor L. Vakaet, head of the Department of Anatomy, University Centre of Antwerp, we owe a great debt of gratitude for teaching us the macroscopic staining method of cartilage, which is the foundation of this study and for introducing us to the methods of embryological investigation.

The consultant surgeons, anaesthesists, radiologists, physicians and pathologists of the Thoracic Unit, Liverpool, England, where we trained in surgery, deserve our sincerest thanks for stimulating our interest in every aspect of bronchopulmonary surgery, physiology, anatomy and pathology. Amongst them we are especially grateful to Mr. F. R. Edwards, M. D., Ch. M. F. R. C. S., for reading our English and advising us on this edition.

Our sincerest thanks are due to Professor Dr. T. H. Schiebler, the Editor and Springer-Verlag for accepting publication of this monograph in the Advances in Anatomy, Embryology and Cell Biology.

References

Aeby, Chr.: Der Bronchialbaum der Säugethiere und der Menschen, nebst Bemerkungen über den Bronchialbaum der Vögel und Reptilien. Leipzig: W. Engelmann 1880

Alescio, T., Cassini, A.: Induction "in vitro" of tracheal buds by pulmonary mesenchyme grafted on tracheal epithelium. J. exp. Zool. 150, 83–94 (1962)

Arey, L. B.: Developmental anatomy, 6th ed. Philadelphia: Saunders 1954

Bariety, M., Paillas, J., Levy, M.: La trachée et les bronches cartilagineuses. Structure et fonctionnement des dispositifs musculaires et élastiques. Paris: Masson & Cie. 1951

Bauhinus, C.: Theatrum anatomicum infinitis locis auctum. Francof. 1621

Benek, G., Endres, O., Becker, H., Nitschke, H.: Wachstum und altersabhängige Strukturveränderungen der menschlichen Trachea. Virchows Arch. path. Anat. 341, 353–364 (1966)

Bidloo, G.: Vindiciae quarandam delineationum anatomicarum contra animadversiones F. Ruijschii praelect. Lugd. Batav. 1697

Boyden, E. A.: Segmental anatomy of the lungs. New York: McGraw Hill 1955

Braus, H.: Anatomie des Menschen. Erster Band: Bewegungsapparat, 2. Aufl. Berlin: Springer 1929

Brenek, I.: Ueber Knorpel- und Drüsenentwicklung im menschlichen Bronchialbaum. Z. mikr.-anat. Forsch. 49, 525–533 (1941)

Brien, P., Dalcq, A.: Caractères généraux des vertébrés. Principes de leur morphologie et de leur évolution. Traité de Zoologie, Tôme XII, p. 3–201. Paris: Masson 1954

Brock, R. C.: The anatomy of the bronchial tree. London: Oxford University Press 1947

Bucher, U., Reid, L.: Development of the intresegmental bronchial tree: the pattern of branching and development of cartilage at various stages of intra-uterine life. Thorax 16, 207–218 (1961)

Buisson, F. R.: In: Bichat, X., Traité d'anatomie descriptive, Tôme IV. Paris: Brosson, Gabon & Cie 1803

Chuang, H., Tseng, M.: An experimental analysis of the determination and differentiation of the mesodermal structures of neurula in urodeles. Sci. Sinica. 6, 669–708 (1956)

Churchill, E. D., Belsey, R.: Segmental pneumonectomy in bronchiectasis. The lingula segment of the left upper lobe. Ann. Surg. 109, 481–499 (1939)

Cordier, G., Cabrol, C.: Territoires pulmonaires, bronches et vaisseaux fonctionnels du poumon. Dessins de Delpech L. Paris: Masson 1959

Cuvier, G.: Leçons d'anatomie comparée. Recueillies et publiées par Dumeril M. 3e éd., Tôme III. Bruxelles: Société typographique belge 1840

Dalcq, A.: L'oeuf et son dynamisme organisateur. Paris: Albin Michel 1941

Dalcq, A.: The concept of physiological competition (Spiegelman) and the interpretation of vertebrate morphogenesis. Exp. Cell Res., Suppl. I, 483–496 (1949)

Daremberg, Ch.: Oeuvres anatomiques, physiologiques et médicales de Galien. Paris: Baillière 1854

Davis, J. D.: Anatomic variations of normal tracheobronchial tree. Arch. Otolaryng. 9, 404 (1929)

Delmas, A., Eralp, I.: Structure de l'éperon bronchique. C. R. Ass. Anat. 41, 1149–1152 (1954)

Diemerbroeck, I. de: Opera Omnia. Meinardum à Dreunen & Giulielmum a Walcheren 1685

Dietzel, K.: Anatomie und Physiologie der Luftröhre und der Bronchien. In: Berendes, J., Link. R. und Zoellner, F., Hals-, Nasen-Ohrenheilkunde. Stuttgart: Thieme 1964

Ewart, W.: The bronchi and pulmonary blood-vessels, their anatomy and nomenclature. With a criticism of Prof. Aeby's views on the bronchial tree of mammalia and man. London: J. & A. Churchill 1889

Fautrez, J.: Rôle des phénomènes d'inhibition comme facteurs de la morphogénèse. Arch. Biol. (Liège) 75, Suppl. 1081–1098 (1964)

Fischel, A.: Entwicklung des Menschen. Berlin: Springer 1929

Foster-Carter, A. F.: Recent work on the anatomy of the bronchi. Brit. Med. Bull. **2**, 245–246 (1944)

Gegenbaur, C.: Traité d'anatomie humaine. Traduit de la 3e éd. allemande par Julin Ch. Paris: Reinwald 1889

Hamilton, W. J., Boyd, J. D., Mossman, H. W.: Human embryology. Prenatal development of form and function, 2nd ed. Cambridge: W. Heffer & Sons Ltd. 1952

Hartmann, R.: Handbuch der Anatomie des Menschen. Strassburg: R. Schulz und Co. 1881

Hayek, H. von: Die menschliche Lunge. Ergebn. Anat. Entwickl.-Gesch. **35** (1951)

Hayek, H. von: The human lung. English translation by Krahl, V. E. New York: Hafner Publ. Co. 1960

Hayward, J., Reid, L.: The cartilage of the intrapulmonary bronchi in normal lungs, in bronchiectasis and in massive collapse. Thorax **7**, 98–110 (1952)

Heiss, R.: Zur Entwicklung und Anatomie der menschlichen Lunge. Arch. Anat. Entwicklungsgesch. Anat. Abt. **1**, 1–130 (1919)

Heller, R., Schroetter, H. von: Die Carina trachea. Ein Beitrag zur Kenntnis der Bifurcation der Luftröhre nebst vergleichend anatomischen Bemerkungen über den Bau derselben. Denkschr. Kais. Akad. Wissensch. Math.-Naturwiss. Cl. **65**, 397–438 (1897a)

Heller, R., Schroetter, H. von: Die Carina tracheae. Ein Beitrag zur Kenntnis der Bifurcation der Luftröhre und ihre klinische Wichtigkeit. Z. klin. Med. **32**, 211–222 (1897b)

Henle, J.: Handbuch der systematischen Anatomie des Menschen. Band II: Handbuch der Eingeweidelehre des Menschen. Braunschweig: Vieweg und Sohn 1866

His, W.: Zur Bildungsgeschichte der Lungen beim menschlichen Embryo. Arch. Anat. Entwicklungsgesch. Anat. Abt. 89–106 (1887)

Hörstadius, S.: Ueber die Folgen von Chorda-Extirpation an späten Gastrulae und Neurulae von Amblystoma punctatum. Acta Zool. **25**, 75–87 (1944)

Horner, W. E.: A treatise on special and general anatomy. Philadelphia 1839

Horsfield, K., Cumming, G.: Morphology of the bronchial tree in man. J. appl. Physiol. **24**, 373–383 (1968)

Huizinga, E.: Ueber den Bau des Bronchialbaumes. Z. Hals-, Nas.- u. Ohrenheilk. **43**, 141–149 (1937)

Huntington, G. S.: A critique of the theories of pulmonary evolution in the mammalia. Amer. J. Anat. **27**, 99–201 (1920)

Hyrtl, J.: Die Corrosions-Anatomie und ihre Ergebnisse. Wien: Braumüller 1873

Hyrtl, J.: Lehrbuch der Anatomie des Menschen, 18. Aufl. Wien: Braumüller 1885

Ihle, J. E. W., Kampen, P. N. van, Nierstrasz, H. F., Versluys, J.: Leerboek der vergelijkende ontleedkunde van de vertebraten: Deel I. A. Oosthoek 1924

Jackson Chevalier, L., Huber, J. F.: Correlated applied anatomy of the bronchial tree and lungs with a system of nomenclature. Dis. Chest **9**, 319–326 (1943)

Kassay, D.: Comments on some controversial questions of bronchial classification and nomenclature. Dis. Chest **40**, 364–373 (1961)

Killian, G.: La bronchoscopie. Traduit par Jankelevitch, S. Paris: Doin 1902

King, J.: On the forms of cartilages which keep open the principal divisions of the bronchial tree. Guy's Hosp. Rep. **5**, 237–241 (1840)

Kitchin, I.: The effects of notochondrectomy in Amblystoma mexicanum. J. exp. Zool. **112**, 393–416 (1949)

Krahl, V. E.: Anatomy of the mammalian lung. Handbook of physiology, sect. 3: Respiration. vol. 1, chap. 6. Washington D. C.: Amer. Physiol. Soc. 1964

Kramer, R., Glass, A.: The bronchoscopic localization of lung abscess. Ann. Otol. (St. Louis) **41**, 1210–1220 (1932)

Lanot, R.: Analyse expérimentale de la formation des somites mésodermiques chez l'embryon de poulet. C. R. Ass. Anat. **52**, 740–747 (1967)

Levy, M.: La trachée et les bronches cartilagineuses. Paris: Thèse 1951

Lieutaud, M.: Anatomie historique et pratique. Paris: Vincent d'Houry et Didot 1777

Lucien, P., Beau, A.: La systématisation pulmonaire. Ses bases morphologiques et ses modalités. C. R. Ass. Anat. **37**, 3–55 (1951)

Lucien, M., Weber, P.: Le système parabronchique externe du poumon humain. C. R. Ass. Anat. **28**, 427–435 (1933)

Lundvall, H.: Ueber Demonstration embryonaler Knorpelskelette. Anat. Anz. **25**, 219–222 (1904)

Lundvall, H.: Weiteres über Demonstration embryonaler Skelette. Anat. Anz. **27**, 520–523 (1905)

Lundvall, H.: Ueber Skelettfärbung und Aufhellung. Anat. Anz. **40**, 639–646 (1912)

Lundvall, H.: Färbung des Skelettes in durchsichtigen Weichteilen. Anat. Anz. **62**, 353–373 (1927)

Luschka, H. von: Der Bandapparat der Santorinischen Knorpel des menschlichen Kehlkopfes. Z. ration. Med. **11**, 132 (1861)

Luschka, H. von: Die Anatomie der Brust des Menschen. Tübingen: Laupp & Siebeck 1863

Marcus, H.: Lungen. In: Bolk, L., Goeppert, E., Kallius, E. und Lubosch, W., Handbuch der vergleichenden Anatomie der Wirbeltiere, Bd. III, S. 909–988. Berlin: Urban & Schwarzenberg 1937

Meckel, J. F.: Handbuch der menschlichen Anatomie, Bd. IV. Halle: Buchhandlung des Hallischen Waisenhauses 1820

Meckel, J. F.: System der vergleichenden Anatomie. Vergleichende Anatomie der Athmungs- und Stimmwerkzeuge, Theil VI. Halle: Verlag der Buchhandlung des Waisenhauses 1833

Menkes, B., Miclea, C., Elias, St., Deleanu, M.: Cercetari asupra formarii organelor axiale. I. Studii asupra differentierii somitelor. Stud. Cercet. sti. Med. **8**, 7–33 (1861)

Merkel, F.: Atmungsorgane. In: Bardeleben, K. von, Handbuch der Anatomie des Menschen, Bd. VI. Jena: G. Fischer 1902

Miller, C. H.: Demonstration of the cartilaginous skeleton in mammalian fetusses. Anat. Rec. **20**, 415–419 (1921)

Miller, W. S.: Thomas Willys and his De Phtisi Pulmonari. Amer. Rev. Tuberc. **5**, 934–949 (1922)

Miller, W. S.: The lung. Springfield: Thomas 1937

Moreira da Rocha, J.: Staining of adult cartilage by Lundvall's method. Anat. Rec. **13**, 447–449 (1917)

Morel, Ch., Duval, M.: Manuel de l'anatomiste. Anatomie descriptive et dissection. Paris: Asselin et Cie 1883

Morgagni, J. B.: Adversaria anatomica omnia. Jos. Cominis, Patavii. 1706

Muchmore, W.: Differentiation of the trunk mesoderm in Amblystoma maculatum. J. exp. Zool. **118**, 137–186 (1951)

Narath, A.: Vergleichende Anatomie des Bronchialbaumes. Anat. Anz., Ergänzungsheft **7**, 168–175 (1892)

Narath, A.: Der Bronchialbaum der Säugethiere und des Menschen. Bibliotheca medica. Abt. A. Heft 3. Stuttgart: 1901

Neil, J. H., Gwynne, F. J., Main, W. W., Fairclough, W. A.: Anatomy of the bronchial tree and its clinical application. Ann. Otol. (St. Louis) **46**, 338–350 (1937)

Nelson, H. P.: The tracheo-bronchial lymphatic glands. J. Anat. (Lond.) **66**, 228–241 (1931)

Palfin, J.: Anatomie du corps humain, avec des remarques utiles aux chirurgiens dans la pratique de leurs Opérations. Paris: Cavalier 1726

Palfyn, J.: Heelkonstige Ontleeding van 's Menschen Lighaem. Leyden: Jan van den Deynster 1718

Patten, B. M.: Human embryology. Philadelphia: Blakiston Co. 1948

Paturet, G.: Traité d'anatomie humaine, Tôme II. Paris: Masson 1958

Policard, A.: Le poumon. Paris: Masson 1955

Quain, J.: The elements of anatomy. London: Longmans Green & Co. 1896

Ragozina, M. N.: The influence of the neural plate and the cord on the development of the axial mesoderm in amphibians. Dokl. Akad. Nauk. SSSR. **51**, 245–247 (1946)

Rap, A. A., Smelt, G. J.: Anatomie en lipiodolonderzoek van de bronchiaalboom. Assen: Van Gorum Co. 1947

Rauber, A., Kopsch, Fr.: Lehrbuch und Atlas der Anatomie des Menschen, Bd. II, 19. Aufl. Stuttgart: Thieme 1955

Reisseisen, F. D.: Ueber den Bau der Lungen. De Fabrica Pulmonum Commitatio. Berlin: A. Rücker 1822

Ruysch, F.: Delucidatio valvularum in vasis lymphaticis et lacteis. Hago Comitis 1665

Ruysch, F.: Responsio ad Godefridi Bidloi libellum quem vindicias inscripsit. Amsterdam 1697

Sappey, Ph. C.: Traité d'anatomie descriptive, Tôme 4, 2e éd. Paris: Adrien Delahaye 1874

Schulze, O.: Ueber Herstellung und Conservierung durchsichtiger Embryonen zum Studium der Skeletbildung. Anat. Anz. 13, 3–5 (1897)

Sicard, J. A., Forrestier, J.: Diagnostic et thérapeutique par le lipiodol. Paris: Masson 1928

Smyth, N. P. D.: The anatomy of the human bronchial tree and pulmonary blood-vessels. Irish J. med. Sci. 6, 269–290 (1949)

Sobotta, J., Becher, H.: Atlas der Anatomie des Menschen, Teil II, 16. Aufl. München: Urban & Schwarzenberg 1965

Sorokin, S.: Recent work on developing lungs. In: de Haan, R. L. and Ursprung, H., Organogenesis. New York: Holt Rinehart & Winston 1965

Spiegelman, S.: Physiological competition as a regulatory mechanism in morphogenesis. Quart. Rev. Biol. 20, 121–146 (1945)

Stefko, W.: Vergleichende mikroskopische Anatomie der Lungen einzelner Mongolen. Z. Anat. Entwickl.-Gesch. 96, 54–67 (1931)

Streeter, G. L.: Weight, sitting height, head size, foot length and menstrual age of the human embryo. Carnegie Inst. Wash. Publ. No 274. Contrib. Embryol. 11, 143–170 (1920)

Streeter, G. L.: Developmental horizons in human embryos. Description of age group XIII, embryos about 4 to 5 long, and age group XIV, period of identation of the lens vesicle. Carnegie Inst. Wash. Publ. 557. Contrib. Embryol. No 199. 31, 27–63 (1945)

Streeter, G. L.: Developmental horizons in human embryos. Description of age groups XV, XVI, XVII and XVIII being the third issue of a survey of the Carnegie collection. Carnegie Inst. Wash. Publ. 575. Contrib. Embryol. No 211. 32, 133–203 (1948)

Takaya, H.: Notochondral influence upon the differentiation and segmentation of muscle tissue. Annot. Zool. Jap. 29, 133–137 (1956)

Vanpeperstraete, F.: Het kraakbeenskelet van de bronchiale boom. Gent: Thesis 1968

Verheyen, Ph.: Ontleedkondige beschryving van het menschen lighaem. Brussel 1711

Vesalius, A.: De humani corporis fabrica libri septem. Basel: Oporinus 1543

Vesalius, A., Valverda: Anatomie of afbeeldinghe van de deelen des Menschlicken lichaems met een Aenwysinge om het selve te ontleden volgens de leringe Galeni, Vesalii, Falopii en Arantii. Amsterdam: Cornelis Danckertz 1647

Weichert, Ch. K.: Anatomy of chordates. New York: McGraw Hill 1958

Wijhe, J. van: A new method of demonstrating cartilaginous microskeletons. Verh. K. Akad. Wetensch., Amsterdam 1902

Willis, Th.: Pharmaceutice rationalis seu diatriba de medicamentorum operatione in corpore humano. Sectio prima: De medicamentis thoracicis. Caput I. De respirationis organis et usu. London 1674

Subject-Index